GLACIER ICE

GLACIER ICE

Austin Post and Edward R. LaChapelle

Revised Edition

UNIVERSITY OF WASHINGTON PRESS *Seattle*

in association with

THE INTERNATIONAL GLACIOLOGICAL SOCIETY *Cambridge, England*

Library of Congress Cataloging-in-Publication Data
Post, Austin.
    Glacier ice / Austin Post and Edward R. LaChapelle. — Rev. ed.
        p.   cm.
    ISBN 0-295-97910-0 (alk. paper)
    1. Glaciers.   2. Glaciers Pictorial works.   I. LaChapelle, Edward R.
    II. title.
    GB2403.2.P68   2000   99-39066
    551.31'2—dc21   CIP

The paper used in this publication meets the minimum requirements
of American National Standard for Information Sciences—Permanence
of Paper for Printed Library Materials, ANSI Z39.48-1984.

Frontispiece photograph: Charles Keeler and Robert Mason examine a
bergschrund at the head of McCall Glacier, Brooks Range, Alaska, 1957

Contents

Illustrations

Preface to the Revised Edition

Not often do authors get a chance to participate in the second edition of a book twenty-seven years after the first edition. *Glacier Ice* went out of print a long time ago, but every year we receive requests for it. From time to time determined purchasers locate a copy in a second-hand bookstore and gleefully report their success to us.

A confluence of circumstances has made a revised edition possible. The authors, both long-since retired, are still active in scientific work and were keen to participate in a revision. The original photographs had been carefully stored: all but two were available for reprinting, and those two have been replaced. The International Glaciological Society of Cambridge, England, in recent years has adopted a policy of reprinting classics in this scientific field and has enthusiastically endorsed this one. We are grateful for the Society's interest and the continuing support of the University of Washington Press.

There have been some changes and additions, although the original illustrations and text remain largely intact. Five photographs, Figures 13, 63, 94a, 94b, and 103, have been replaced with better depictions of the same subject. Several errors have been corrected, and a few changes have been made to update the text. A bibliography has been added to improve the usefulness of the new edition as a reference. Finally, and most obviously, the book's original oversize format has been changed to be more bookshelf friendly.

We thank all the glacier fans whose enthusiasm over the years has kept *Glacier Ice* alive and ready for this timely resurrection.

Austin Post
Edward R. LaChapelle
November 1999

Preface to the Original Edition

In 1955 the first organized aerial photography of glaciers in western North America was started under the auspices of the University of Washington by Richard C. Hubley. His photographic reconnaissance of the North Cascades brought to public attention the extent of glacierization in this area and identified numerous glacier advances generated by a moderate trend toward a cooler and wetter climate. Hubley repeated this survey each year until his untimely death in northern Alaska in October, 1957, during the International Geophysical Year. LaChapelle continued the Cascade Range photo flights through 1960 in order to maintain the record of glacier fluctuations there.

Beginning in 1960, Post expanded the annual photo-survey of glaciers along the north Pacific Coast of North America and into the interior ranges of Alaska. This work was supported by National Science Foundation grants to the University of Washington. In 1964 Post joined the Water Resources Division of the United States Geological Survey. Since that time the photographic reconnaissance of glaciers in western North America has been continued by the Survey and has expanded to include other mountain ranges. The extent and character of glacierization has now been well documented, the patterns of glacier change are beginning to emerge, and the extent of phenomena like glacier surges has been established.

These aerial reconnaissance techniques have now become firmly established as a tool of glaciology. A by-product is the accumulation of a very large file of aerial photographs of glaciers, a file that is augmented by hundreds of new negatives each year. As we have catalogued and studied these photographs, continuing our scientific research on glacier extent and behavior in North America, we have encountered outstanding examples of many well-known (and some not so well-known) glacier phenomena. This suggested at an early stage that material was at hand for an illustrated text on glacier features.

The majority of photographs in this book are taken from Post's aerial photograph files. As some features could best be illustrated by a ground-based camera, we have included numerous other photographs taken in the course of our glacier research projects. Inquiries among our colleagues led to additional illustrations of certain features missing from our own files. The final compilation that appears here is thus gathered from several sources, with Post's aerial photographs as the foundation.

There is a hidden contribution to most of these photographs which is not obvious to the armchair reader, but which we wish to emphasize and acknowledge. Aerial photography of glaciers requires not only a properly equipped photographer—it is a team effort, and the other half of the team is the aircraft pilot. Weather conditions over the mountains of the North Pacific Coast are often poor, and good picture weather is rare. The glacier areas are remote, navigation aids are scarce, and the terrain is as inhospitable to lost fliers as any in the world. Turbulent air and sudden downdrafts over the mountains can treat a light aircraft to unexpected and awesome roller-coaster rides. A high order of piloting skill and a thorough knowledge of the mountains and weather are essential to a safe return. In addition, the pilot must understand and even anticipate the desires of the photographer so that he can position the aircraft for the best pictures without wasting time and fuel. Many glacier photographs taken on the ground also depend on a skillful pilot, for much modern glaciology relies heavily for logistic support on ski-wheel aircraft which can land and take off under adverse weather and snow conditions from tilted and formidably short sections of glaciers.

For both our aerial photographs and our air-supported glacier research projects, the pilot was the same: William R. Fairchild of Port Angeles, Washington. Without his remarkable abilities most of the photographs in this book would never have been taken.

On February 5, 1969, Bill Fairchild was killed in a tragic crash while taking off from the Port Angeles airport.

To the memory of Richard C. Hubley, brilliant glaciologist and originator of aerial glacier studies, and of William R. Fairchild, the pilot whose skill made these projects possible, this book is sincerely dedicated.

Austin Post
Edward R. LaChapelle

GLACIER ICE

1. Crystals of lake ice (candle ice)

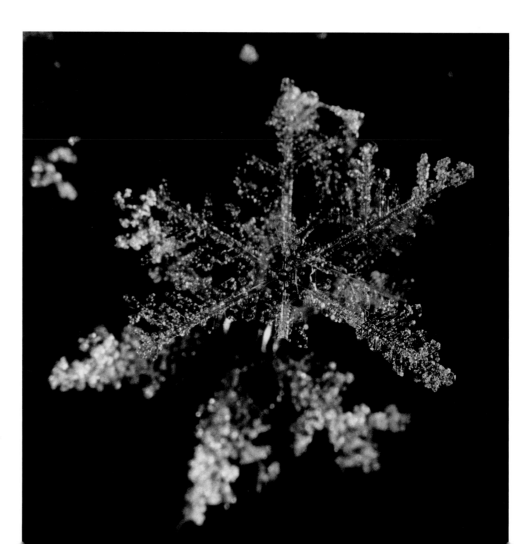

2. Snow crystal

Glacier Formation and Mass Balance

When water freezes in bulk, the shape of the ice crystals formed depends on the rate of freezing and the manner in which the water is exposed to low temperatures. On the surface of the earth, this process commonly takes place on the open surface of oceans, lakes, and ponds exposed to cold air. In fresh water the characteristic *candle ice* crystals of lake ice are formed. These columnar crystals, which form at right angles to the water surface and often reach several centimeters in length, are not readily discernible in the solid sheet of ice covering a lake, but they become visible under favorable conditions when sunlight melts the crystal boundaries and disaggregates the ice into individual crystals. Such typical candle ice is seen in Figure 1.

Water vapor in the atmosphere also freezes, first forming tiny ice crystals which under various conditions grow to become snow crystals of almost infinite variety. A *glacier* is formed on land in areas where more of these snow crystals accumulate annually than melt. Over a period of many years these accumulations become deep enough to form glacier ice, which by a number of processes flows because of its own weight. A glacier is most simply defined as a mass of perennial ice on land that shows evidence of glacier flow.

The crystals of glacier ice thus are not "frozen water" in the sense of lake ice. That is, they did not gain their present form by the freezing of liquid water. Instead, these crystals are the product of extensive metamorphism. Snow crystals like those shown in Figure 2 have a structure entirely different from that ultimately exposed at a glacier surface. The conversion to glacier ice may be aided initially by melting and refreezing in the surface snow layers, but it is the long, slow process of metamorphism, water molecule exchange from one ice particle to another, that eventually over years, decades, and centuries converts the original tiny snow particles into large glacier crystals. The ultimate size of individual crystals is influenced by glacier flow as well as age; under favorable circumstances they may reach 30 centimeters or more in diameter. Melt by the sun also disaggregates the surface of glacier ice, but the crystal shape revealed, such as shown in Figure 3, is entirely different from that of lake ice (Fig. 1).

Each winter a mountain glacier is blanketed with fresh accumulations of snow that may reach depths of as much as 6 to 10 meters in a maritime climate. The thickness of this accumulation usually increases with altitude up to the level of maximum precipitation, which generally occurs at altitudes of 1,000 to 3,000 meters along the north Pacific Coast. In the summer much of this new snow is removed by *ablation* (melting), the rate of melt usually decreasing with altitude. As a result of these altitude variations, the winter snow vanishes first from the glacier terminus, and then from successively higher altitudes on the glacier. The boundary between the winter snow accumulation and the bare ice of the lower glacier exposed by summer melt is called the *transient snowline*, or simply the *snowline* (see Figs. 5 and 35). This should not be confused with the *regional snowline*, which is defined as the altitude at which annual accumulation balances ablation on a ground surface. Accumulated snow and its lower limit, the snowline, persist on a glacier long after snow has vanished from the surrounding rocks, for unbroken snow surfaces are much less sensitive to melt by solar radiation than are those interrupted by rocks and trees.

3. *Crystals of glacier ice*

Snow that survives a year or more of ablation on a glacier is called *firn.* The contact between bare ice and firn is called the *firn edge.* At the end of the summer the snowline reaches some maximum altitude on the glacier. This is the *firn line.* Above it is the *accumulation area,* where more snow is deposited than melts year after year and is slowly compressed into ice. Below the firn line is the *ablation area,* where all annual snow melts most years and ice is also ablated away. Under the influence of gravity, ice flows from the accumulation to the ablation region. A method frequently used to measure a glacier's *annual mass balance* is to determine its *accumulation area ratio* (AAR), the area of the glacier's accumulation divided by the glacier's total area. Under ideal steady-state conditions the snow accumulation, ice flow, and snow and ice melt all balance so that no change in mass occurs. As such conditions are rare in nature, glaciers are constantly adjusting to changes in snow accumulation and ablation by increasing or decreasing their flow velocity and by advance or retreat of the *terminus,* or lowest point of the glacier.

Snow accumulation on a glacier is irregular, being much thicker one place than another even at the same altitude. The

4. *Wind-controlled snow accumulation, unnamed glacier,*
Kenai Mountains

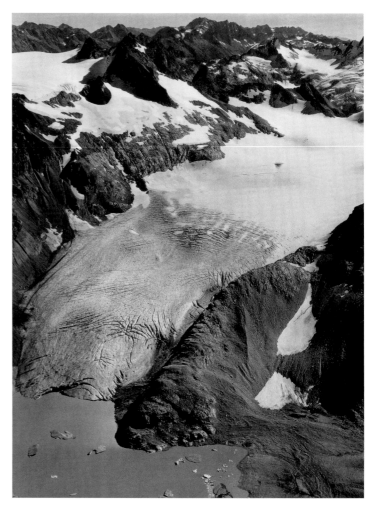

5. *Snowline, South Cascade Glacier, North Cascades, 1957*

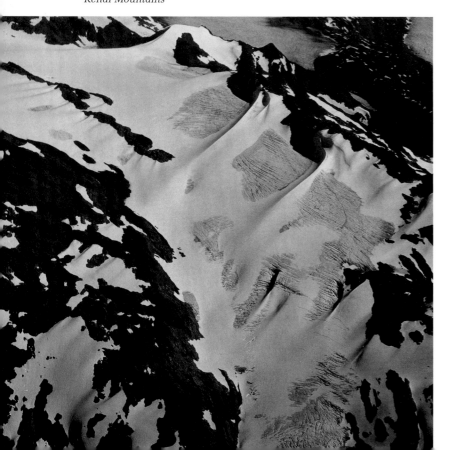

relation of local topography to prevailing winter storm winds determines which part of the glacier will receive the deepest snow deposition. Figure 4 shows an extreme example of terrain effects on wind-drifted snow. Summer ablation also takes place unevenly, for it is affected by the relation of local topography to sun and warm winds. The variations of accumulation and melt do not coincide. These differences are expressed by the snowline and firn line mentioned above. The example illustrated in Figure 5 is a simple case; often the firn edge is highly irregular when the ablation surface intersects uneven accumulation horizons. When exceptionally heavy summer melt occurs, more than one season's accumulation may be intersected. In the North Cascades severe melting occurred in 1958, and the South

6. *Firn edges, South Cascade Glacier, North Cascades, 1958*

7. *Tiger Ice Cap, Baffin Island*

Cascade Glacier presented a strikingly different firn edge picture in Figure 6. Snow layers from six different years of accumulation were exposed when this photograph was taken.

A similar feature in an entirely different climate appears in Figure 7. These small, stagnant glaciers are part of the Tiger Ice Cap, Baffin Island. Annual accumulation here is very small compared with the Cascade Range of Washington. At such a high latitude (more than 70° N), ablation is also much lighter.

But a similar pattern of firn edge has developed, probably for the same reason as on the South Cascade Glacier. The surrounding area exhibits *soil polygons,* a type of patterned ground common to permafrost areas at high latitudes. The identification of each annual firn layer in such a picture is based on the dust and dirt that accumulate on the glacier each year.

Such identification is not possible in very clean, dirt-free snow, such as that found on the dust-free central ice sheets of

Greenland and Antarctica. But in alpine regions the wind picks up dust from both near and distant bare ground and deposits some of this along with snow on the glacier. Such dust is thinly distributed throughout the winter snowpack and is not visible in winter or early spring when the snow presents a clean white surface. As spring and summer melt proceed, this dirt and dust accumulate on the surface. An *ablation surface* thus develops, with its characteristic surface dirt layer, which becomes darker as summer progresses. Such an ablation-formed dirt horizon divides one year's firn accumulation from the next. When melt is strong enough to remove entirely the annual snow, its dirt layer is added to the next layer underneath. If ablation continues, this process can result in the dirt from several years' accumulation collecting together, each addition giving the exposed surface a successively darker color as shown in Figure 8. The many shades of surface "dirtiness" (seen in Fig. 6) have been formed in just this fashion, and the pattern of their boundaries reveals clearly the characteristic distribution of snow accumulation on this glacier.

In many ways the formation of glacier ice is analogous to the formation of metamorphic rock structures in the earth's crust. (From the geological viewpoint, in fact, ice is a rock—simply one with a much lower melting point than most.) Instead of lime, silt, and other rock materials being deposited by sedimentation on a lake or ocean bottom, a layer of snow is deposited by sedimentation from the atmosphere. Successive layers compress those beneath, and metamorphism, influenced by percolating meltwater and prevailing temperatures, gradually converts the snow crystals to new forms, just as it does rock particles. The stratigraphy, or layering, in firn and ice may remain visible for many years wherever a section is exposed by fracturing or erosion. Glacier flow first distorts this layering and then completely obscures it, although visible layers will persist a long time if the glacier motion is simple and unbroken. The layering seen in Figure 9 has been tilted as it was carried down to the zone of melt, but is still clearly recognizable. This can happen only if the course of the glacier is free of ice cascades or other irregularities and little shearing takes place.

Accumulation layers are also visible in Figures 10 and 11. Such layers, their form distorted by flow and by ablation, can also be recognized in Figure 71. Figure 11 exhibits a complex record of weather influence. Most of the ice in this exposed face was deposited as shallow layers which have been deformed downward by flow at the center to form gentle curves. At the point indicated by the arrow, an abrupt transition in the layering occurs. Identical transitions are found in sedimentary rocks, where they are known as *unconformities.* In the case of both rock and ice, they represent an interruption in the regular,

8. The formation of ablation surfaces

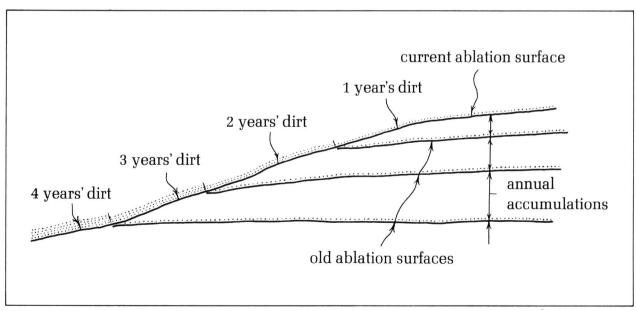

current ablation surface

1 year's dirt

2 years' dirt

3 years' dirt

4 years' dirt

annual accumulations

old ablation surfaces

9. *Accumulation layering in glacier ice, Chakri Glacier, Garwhal Himalayas*

10. *Accumulation layering in firn*

11. Accumulation layering and unconformity in glacier ice, Blue Glacier, Olympic Mountains

12. *Avalanche-fed glaciers, Mount Everest*

annual sequence of accumulation. Exposure for a time to erosion (melting in the case of ice) results in a surface which may not necessarily conform to the deposition surface. Subsequently, when accumulation resumes, layers are deposited whose surfaces conform to the intervening erosion surface. In the case of the Blue Glacier (Fig. 11), recent climatic changes stand clearly revealed. The erosion surface represents a period of intensive glacier melting known to have occurred in the Pacific Northwest during the 1930s and 1940s. Subsequent deposition of the nearly horizontal layering above the arrow occurred during the recent period of glacier rejuvenation beginning in 1948. Ice below the unconformity must therefore have been laid

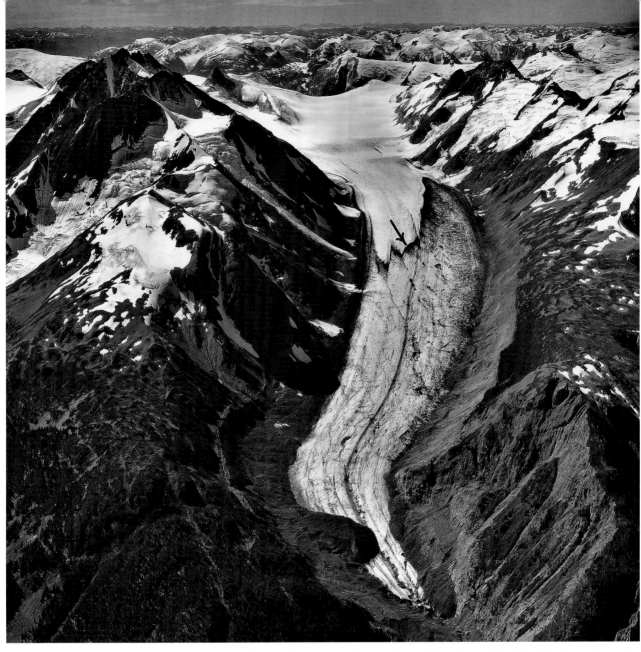

13. Dark ash layer at demarcation between firn and bare ice, British Columbia, Coast Mountains

down as snowfalls before the 1930s. Presumably the layering displayed here is annual, hence the ice at the foot of this face must date from the turn of the century or earlier.

Not all glaciers exhibit a primary layering built up by annual snow deposition from the atmosphere. Some, for instance, consist almost entirely of ice formed by refrozen melt-water that percolated down from surface snow melt. A rather critical combination of annual snow accumulation and seasonal mean temperature variations is required for this. Others are

nourished mostly by snow avalanches sliding from surrounding steep peaks, a form of secondary deposition which accumulates a jumbled mass of snow with much less distinct layers. Figure 12 illustrates terrain typical of avalanche-fed glaciers.

Glacier ice, like rock, preserves in its layering the history of weather influences that went into its making. Sometimes unusual events will leave a special mark in the stratigraphic profile; such an example is indicated by the arrow in Figure 13. This small, unnamed glacier in the British Columbia Coast

14. Ash layer preserved in ice stratigraphy, unnamed glacier in Harding Ice Field, Kenai Mountains

Mountains exhibits a striking black line delineating the edge of the retreating snow line. This is the remnant of a volcanic ash layer deposited on the glacier. Meltwater has washed away the ash from the bare ice below the snow line. Residual winter snow blankets the firn, within which the accumulated ash has been concealed. Only at the retreating snow and firn edge is the ash exposed to view. Note also that more than one rockfall has overridden parts of the bare ice near the terminus.

A similar ash layer of earlier origin (probably the Mount Katmai eruption of 1912) appears in Figure 14, which shows an outlet glacier draining the Harding Ice Field. This layer was incorporated into the accumulation zone long ago and has been transported by glacier flow almost to the terminus, where extensive ablation has diffused the ash over the underlying ice. The upper ash boundary is still sharp where it has been protected by the layers of ice on top of it.

The annual mass balance of a glacier is the difference between the amount of ice it gains each year (usually as snowfall) and the amount it loses (usually as melt). In order for a glacier to grow, it must have a sustained positive mass balance (net mass gain each year). If the mass balance is persistently negative, the glacier will shrink. The relations between a glacier's mass balance and its terminal advance or retreat are complex. Considerable time lag may exist before the one has an effect on the other. The state of health of a glacier (the character of its mass balance) is not always obvious from the behavior of its terminus. The position of the snowline is a better indicator, but it does not by any means tell all the story. An accurate assessment of a glacier mass balance can be gained only by actually measuring the ablation and accumulation. Part of the accumulation-measuring process, determining the water content of snow, is shown in Figure 15.

When does a snowfield reach a sufficient size to become a glacier? Or, conversely, when does a retreating glacier cease to be one? To qualify as a glacier, a body of ice ought to have at least some true glacier ice, whose pore volume is no longer communicating like that of snow. It ought to exhibit glacier flow, but this immediately raises a problem, for even a winter snow cover is subject to rapid deformation under the influence of gravity. Ice motion exhibited by at least one *crevasse* somewhere on the glacier is a requirement demanded by many glaciologists. The geologist, on the other hand, may point to geomorphological features, the effects of ice on the landscape such as *moraines* or *striae* (see Figs. 45 and 105), as the positive indications of a glacier flow and hence a real glacier.

An example of a marginal glacier appears in Figure 16. The snow patch might be a perennial feature, or just one that happened to survive *this* particular ablation season. Old snow or ice could be hidden underneath the mantle of rock debris. Insignificant as the snow patch in Figure 16 may be, there still is no disregarding that large, obvious *terminal moraine* beneath it. In fact, it is that moraine which raises the problem of identification in the first place; without it this tiny ice mass would be considered to be no more than a lingering snow patch.

15. *Accumulation survey,
Blue Glacier, Olympic
Mountains*

16. *Unnamed glacier,
Wasatch Mountains*

Major Surface Features of Ice

At the head of most alpine glaciers (as distinguished from ice sheets) there is a zone where snow accumulation rapidly diminishes with altitude, often where the glacier approaches a wind-swept ridge or a rock wall too steep to support an ice mantle. Within this zone active flow is not possible, for too little snow accumulates each year. Below this zone the normal process of mass accumulation and flow sustains an appreciable glacier motion. A persistent feature of most active glaciers is the *bergschrund,* a large crevasse which separates the flowing ice from the nearly stagnant ice mass above. Often it runs clear across the head of the glacier, where it constitutes a late-summer obstacle to mountaineers when snow bridges have melted away.

The bergschrund should be distinguished from another glacier feature whose name also originated in the Alps. This is a *randkluft,* the space or crack between the edge of the glacier and the adjacent rock wall. It is bounded on one side by ice and on the other by rock, while the bergschrund divides two ice masses. The randkluft is developed predominantly by melt adjacent to the dark rock, rather than by flow.

Bergschrunds are visible in many illustrations in this book. Several are especially clear in Figure 17, where they are indicated by arrows. Above the right-hand arrow the stagnant ice mantle has thickened until it has begun to flow and form new cracks above the present bergschrund. Eventually these new cracks will became the bergschrund, while the old one below closes and is carried down the glacier.

Figure 18 provides another illustration of glacier accumulation areas. The basin in the middle of this picture shows by its heavily crevassed and broken surface that it is in rapid motion and discharging a large amount of ice each year by vigorous flow. The quantity of ice any glacier must transport by flow each year from a higher to lower altitude depends on the rate at which mass is added by snowfall above and removed by melt below. The *activity index* of a glacier is the rate of increase of net mass balance with altitude. The larger the activity index,

the greater is the annual volume of ice that must be transported across the firn line by glacier flow. Active glaciers with vigorous flow are most common in maritime climates where snow lies deep in the high mountains and summer melt is strong at lower elevations. Glaciers in a continental climate usually have a much lower activity index.

Seen from the ground, the broken surface of an accumulation area such as that in Figure 18 is a chaotic jumble of crevasses and *seracs,* or isolated ice blocks, which seriously impede travel by mountain climbers. Passage through such a zone may be possible early in the melt season while deep snow still blankets the obstacles (Fig. 10), but as summer melt erodes the blocks and uncovers the crevasses, travel becomes increasingly difficult (Fig. 19).

To many people, a glacier means ice—bare, blue ice. This expected aspect of a glacier is found only below the firn line in the ablation zone. Here the summer melt removes entirely, each year, both the winter snow and some of the ice that has been formed by compaction in the accumulation zone and carried down by glacier flow. In extensive high mountain areas where annual snow accumulation far exceeds melt, the glaciers from many accumulation areas covering hundreds of square kilometers may coalesce to form a single great river of ice as their common ablation zone.

The Bering Glacier in Alaska (Fig. 20) is an outstanding example of such an ice river. The part shown in the photograph was taken where the Bering Glacier is 11 to 16 kilometers wide. Covering an area of over 5,100 square kilometers, this glacier system is the world's largest outside of Greenland and Antarctica.

Because the ablation zone surface is not buried by snow accumulation throughout each year, it can display clearly the crevasse patterns and other features that persist from year to year. The lower reach of the Sumquolt Glacier shown in Figure 21, which has recently experienced retreat, shows a strong fracture pattern of crevasses generated by flow of the ice around a bend in its channel (upper center). To the left of this pattern is a fainter one with fractures at a different angle, formed by a flow feature farther upstream. Where these two patterns

17. *Bergschrunds, British Columbia, Coast Mountains*

19. Seracs in firn, Blue Glacier, Olympic Mountains

18. (opposite) Accumulation zone crevasses, Bishop Glacier, British Columbia, Coast Mountains intersect (just left of center), a field of seracs is formed. Seen from the ground such ice features are similar to the firn blocks of the accumulation zone, but they consist of solid ice which, as shown in Figure 22, is often eroded and rounded by melt.

Characteristic crevasse patterns are produced by shearing in the glacier ice induced by friction of the valley walls between which the glacier flows. The patterns are different in the accumulation and ablation zones. In the accumulation zone, the glacier flow velocity increases downstream because the amount of accumulated mass that must be transported also increases downstream. Such a phenomenon is called *extending flow.* Combined with side friction, it produces crevasse patterns like the one shown in Figure 79. The reverse situation is found in the ablation zone. Here the amount of mass that must be transported decreases downglacier, and so does the flow velocity. This is called *compressing flow,* which combines with side friction to produce *chevron crevasses* like those seen along the margins of the Sumquolt Glacier in Figure 21.

The Sumquolt Glacier is relatively slow-moving, with an ablation zone exposed to substantial melt each summer. These factors combine to produce a blurred pattern of seracs and crevasses which does not exhibit all the jagged angularity of freshly fractured ice. On the other hand, in Figure 23 the surging Gakona Glacier displays a fresh and unobscured pattern of fractured glacier ice. Here the complex strains in the rapidly flowing ice have shattered the glacier surface into orderly arrays of seracs. These seracs have suffered very little melt and have maintained their original form and orientation for some distance downglacier from the point where the ice surface was broken apart. A glacier surface like this is almost impossible to cross on foot.

An *icefall* demonstrates very conspicuous evidence of active glacier flow, and its broken, chaotic surface may be impossible for the most expert mountaineer to traverse. Flow velocity in an icefall may reach a thousand or more meters per year, much faster than in other parts of most glaciers. In spite of the name, this feature is still not an ice "fall" in the sense of an avalanche of ice from the terminal of a *hanging glacier* which cascades down a mountainside in seconds or minutes. The flow of the underlying ice is accelerated by the steep gradient and constricted channel, while the surface layer of the glacier, unable to accommodate to this change, is broken into crevasses and seracs which are pulled apart by the extending flow at the top. Many of the individual seracs and ice blocks do collapse and tumble down, one by one. At the foot of the

icefall the motion decelerates, and compressing flow rapidly closes the fissures and restores a relatively smooth ice surface.

Icefalls frequently separate plateaus or basins, which serve as accumulation zones, from the valley ablation areas of mountain glaciers. The series of steep ice cascades in Figure 24 is an extreme example. Another spectacular example is the icefall of the Margerie Glacier in Figure 25, a steep zone in a valley glacier where it descends from the generating firn fields located on a large, upland plateau.

From the ground, icefalls appear just as rough as they do from the air. Normally they are avoided as a route of mountain ascent, both because they are difficult to travel and because of danger from falling seracs. Occasionally no alternate route is available and climbers are forced to travel paths like the famed Khumbu Icefall on Mount Everest (Fig. 26).

20. Bering Glacier, south central Alaska

21. *Crevasse patterns, Sumquolt Glacier, British Columbia, Coast Mountains*

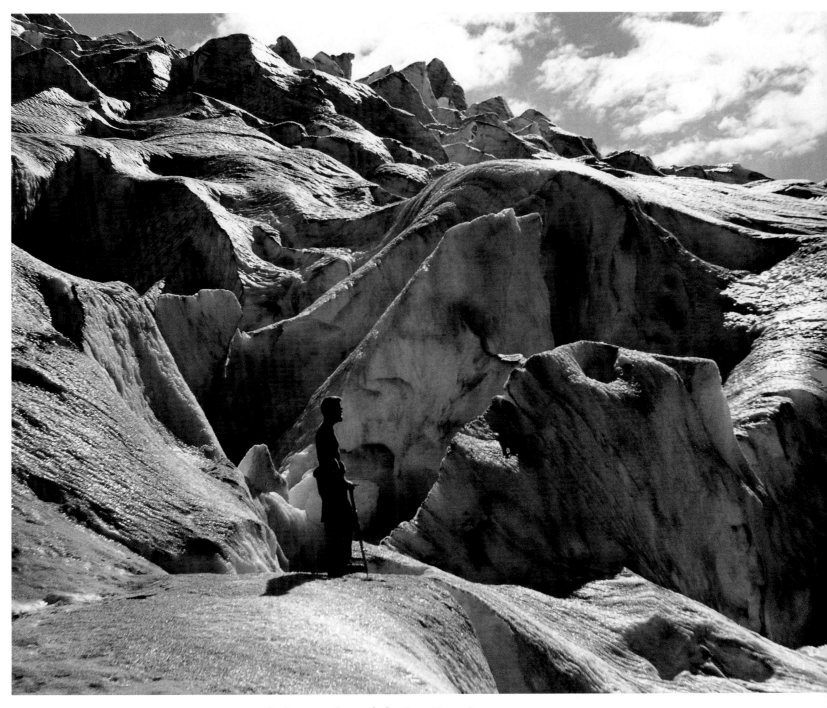

22. *Broken ice and seracs, ablation zone, Lemon Creek Glacier, southeast Alaska, Coast Mountains*

24. Icefalls, Mount Ratz, British Columbia, Coast Mountains

23. (opposite) Crevasse patterns resulting from very rapid surge, Gakona Glacier, Alaska Range

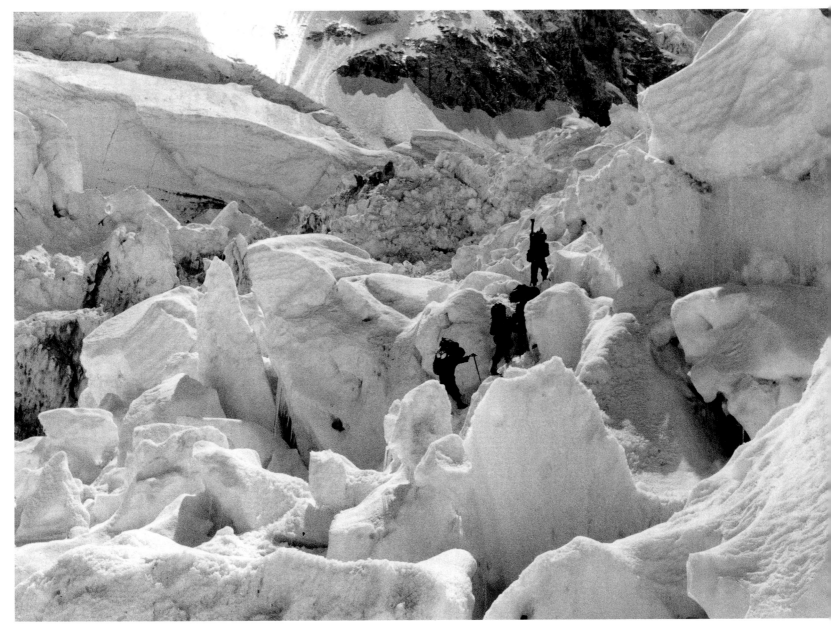

26. *Khumbu Icefall, Mount Everest*

25. *(opposite) Icefall, Margerie Glacier, Fairweather Range, Glacier Bay*

Glacier Flow

In order to produce the characteristics of glaciers—mass transfer, crevasses, erosion, moraine formation—the ice must flow. Taken from the simplest viewpoint, glacier ice is a substance that flows down the mountainside much as a stream of water does, but at a far slower rate. The plastic qualities of ice can be measured in the laboratory, but these are difficult to apply in a simple fashion to calculate glacier behavior. The large-scale motion of a temperate glacier (at the melting point throughout) depends on a complex combination of slip along the bed, fracturing, and ordinary plastic deformation.

Glacier motion also appears to depend to some extent on the amount of meltwater present within and under the glacier. Some of these complexities of motion are suggested in Figure 27a, b, and c. The changes over a period of 2 1/2 months are due not only to flow but to collapse of seracs and ablation. Analysis of the complete series of time-lapse photographs from which these are selected shows that individual ice blocks flow sideways as well as down the icefall. The motion is so erratic that some points actually descend in a zigzag fashion.

At the bottom of a glacier there is a layer of dirty, debris-

27. Time-lapse sequence of icefall, Blue Glacier, Olympic Mountains   *a*

laden ice called the *glacier sole.* Rocks of all sizes embedded in this layer provide the abrasive tools with which the glacier scours and erodes its bed. Pure ice alone would have far less erosive effect on solid and massive bedrock. Deformation of ice in which the rocks are embedded allows it to conform to the shape of the glacier bed and develop a grooved pattern on its under surface.

Access to the bottom of a moving glacier is difficult, and the available photographs of a glacier sole are very few indeed. An aerial photograph is hardly the place where one would expect to find pictured the underside of a glacier, but a fortunate chance occurrence has provided in Figure 28 just such an illustration. The South Sawyer Glacier at the head of Tracy Arm, a long fjord in southeastern Alaska, had just discharged an iceberg at the time this photograph was taken. The right-hand arrow points to an upturned piece of the glacier sole, stri-

ated by the rock bed over which it moved. The left-hand arrow points to another striated ice surface which may have had a different origin, etching due to fresh water from subglacial rivers flowing up the glacier ice cliff to the surface. The photographer was not aware of his luck when he made the exposure; only later did more leisurely examination of the finished print reveal this unusual feature.

The manner in which the grooved undersurface of the glacier sole is generated is seen in Figure 29, another photograph obtained by a chance stroke of luck when glaciologist F. Mueller discovered a cavern in the Pumori South Glacier (near Mount Everest) which gave direct access to the glacier bed. The sliding ice is imprinted with large and small grooves by each bump and irregularity in the bedrock boss at the picture center. A glacier like this slides over the ground by two processes. Plastic deformation of the ice allows it to bend and

b

c

28. *Glacier sole exposed on icebergs, South Sawyer Glacier, southeast Alaska, Coast Mountains*

slide around large obstacles, but it is too stiff to bend around small ones a few centimeters high. Instead, it slides over these by temporarily reverting to liquid water, a process known as *regelation.* The ice melts as a result of pressure on the uphill side of small obstacles, flows around them as water, and refreezes at the downhill side, accompanied by a countercurrent of heat which sustains the process. If a space exists on the downhill side, the refrozen water sometimes forms a feathery deposit of *regelation spicules.* These spicules festoon the upper part of the rock in the center of Figure 29.

29. *Glacier sole and grooves, Pumori South Glacier, Garwhal Himalayas*

At another part of this same cavern the complex motion of the glacier sole occurs both as fracturing and as plastic deformation around a boulder. Figure 30 shows numerous rocks embedded in the moving ice; these provide the glacier with its tools of erosion. The layer of debris-laden, contorted ice that makes up a glacier sole is seldom as clearly exposed to the light of day as it is along the edge of the Pumori Glacier as shown in Figure 31. The origin of the fold at right is obscure, but this feature offers unmistakable evidence that the process of a glacier sliding over its bed is far from a simple one. The sharp demarcation of the dark, debris-laden sole from the clean glacier ice above is normal.

30. *Glacier sole with deformation and englacial debris, Pumori South Glacier, Garwhal Himalayas*

31. Glacier sole, folding, and deformation, Pumori South Glacier, Garwhal Himalayas

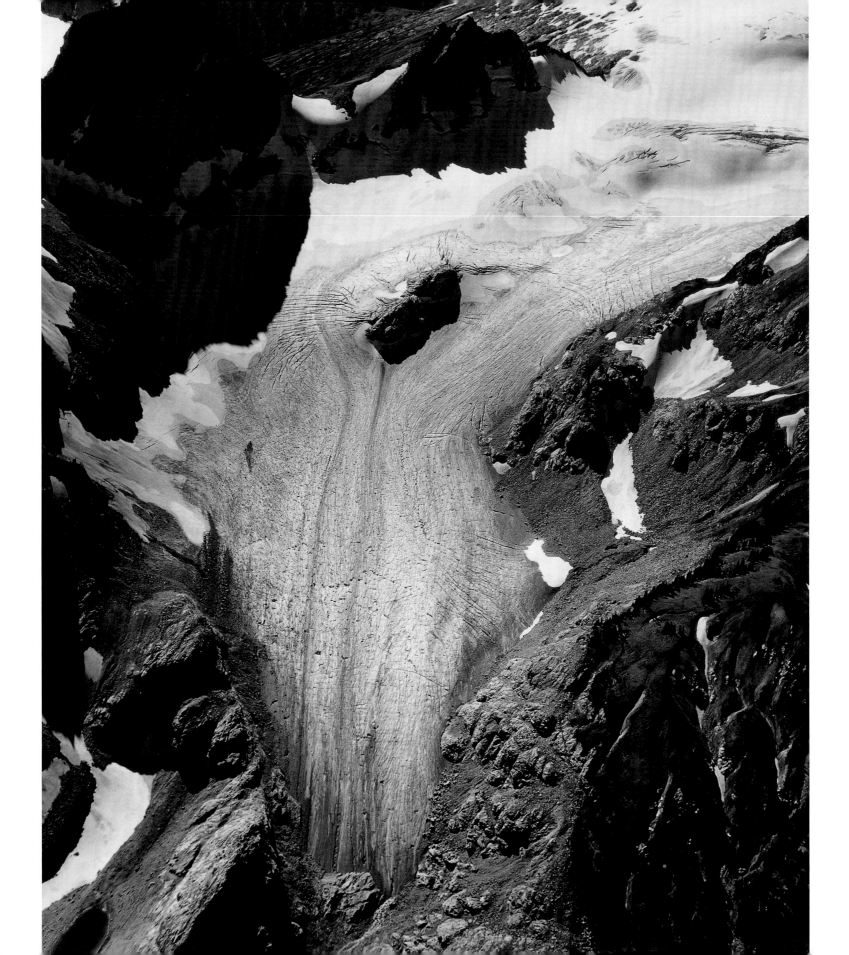

Glacier Fluctuations

Glaciers are not a fixed feature of mountain landscapes. They wax and wane with changes of climate; even in a stable climate their length may oscillate. For the past century, most mountain glaciers of the world have receded, although local interruptions of the general retreat have appeared from time to time. In a retreating glacier, lower reaches are stagnant or flow too sluggishly to replace the ice melted away each summer. The ice shown in Figure 32 slopes gently at the terminus, its surface broken by few crevasses. If the retreat has proceeded far enough, the valley tongue disappears entirely. Figure 33 shows such a decaying glacier, its icefall only a small remnant that reveals a band of bedrock cliffs over which it formerly flowed with enough vigor to fill the valley floor with ice.

The powerful erosive force of an advancing glacier scours soil and bedrock free of vegetation; when the ice retreats, an abrupt transition from bare, mineral rock to mature plant communities is left behind. In Figure 34 this "high ice" mark, called the *trimline,* is especially clear along the margins of Franklin Glacier and its tributaries.

32. (opposite) Retreating glacier tongue, Honeycomb Glacier, North Cascade Range

33. Desiccated icefall, Coronet Peak, Canadian Rocky Mountains

34. *Trimlines, Franklin Glacier, British Columbia, Coast Mountains*

35. *(opposite) Trimlines, tarn, and moraines, Pattullo Glacier, British Columbia, Coast Mountains*

36. *Advancing glacier margin, Taku Glacier, southeast Alaska, Coast Mountains*

37. Retreating and advancing glaciers, unnamed glaciers near Alaska-Yukon Border, St. Elias Mountains

The eroded ground and mantle of debris exposed in the former bed of a retreating glacier is a desolate landscape, at first broken only by *tarns,* small lakes occupying hollows scooped out of the terrain by the glacier flow. At lower altitudes, plants soon appear as the forest begins to re-establish itself on the barren ground, gradually covering the scar and obscuring the trimline (Fig. 35).

Sustained glacier recession in most parts of the world today has accustomed mountaineers and glaciologists to the zone of bare, eroded earth between vegetated ground surfaces and glacier ice (Figs. 34 and 35). These denuded zones are covered with an unstable rock and dirt mantle which is often unpleasant to traverse. It comes as a refreshing change to a footsore mountaineer when he can step directly from clean, blue ice into a mature forest. Such an invasion of timber by advanc-ing ice is rare today, though it must have been common two or three hundred years ago. Many cases of glacier advance are currently being recorded, but most of these have not yet recovered ground lost in recessions of the recent past, so the ice is still surrounded by barren moraines. The Taku Glacier is an outstanding exception. It has advanced steadily for the past fifty years, and today continues the destruction of forests along its margins as shown in Figure 36.

The steep, bulging terminus and a heavily crevassed tongue are both marks of an advancing glacier. The contrast with a receding one is especially clear in Figure 37, where two glaciers of similar shape and size, lying side by side, are in opposite states of behavior. This is evidently due to surging; the left glacier is in its passive phase and the one on the right is in its active phase.

38. *Different stages of advance and retreat, "Three Congruent Glaciers," St. Elias Mountains*

The glaciers in Figure 38, in Yukon Territory, have gained the unofficial name of the "Three Congruent Glaciers." They originate from very similar firn basins and flow along nearly identical parallel valleys. The one on the left has recently advanced. Its thick, bulging terminus overrides debris in the valley floor, while the thickening edges accumulate a rock mantle from the adjacent valley walls. The center glacier also advanced, but now recedes and has been doing so for a number of years. The rock mantle along each side has been deposited as *lateral moraines,* ridges of unconsolidated material parallel to the glacier flanks. The third "Congruent Glacier" appears to have made a very recent advance from which it has not yet begun to retreat. Faint trimlines considerably higher up the valley walls than the present ice levels indicate that all three glaciers have at one time advanced much farther than their positions today.

In Figure 39, the surging north branch of the Tyeen Glacier in the Fairweather Range has advanced rapidly out of its accumulation basin and onto the surface of the main glacier. The exceptionally broken and crevassed surface indicates very rapid ice motion.

A peculiar type of glacier activity is illustrated in Figure 40a and b. The steep firn field in Sunset Amphitheater, the major accumulation zone of the South Mowich Glacier on Mount Rainier, discharged a vigorous flow of ice in 1960, as shown in Figure 40a. In fact, this steep area of the glacier had turned into an active icefall right up to the bergschrund. Badly broken ice flowed through the narrow breach in a cliff band onto the lower glacier. Two years later, as shown in Figure 40b, the situation had changed drastically: the firn field was now quiescent, and the ice flow through the breach nearly dried up. The South Mowich Glacier exhibits behavior not uncommon to many very steep glaciers. They tend to build up accumulation, reach a critical level, discharge a sudden flow of ice, then return to a quiescent state while accumulation resumes again. The "Spillway Glacier" in the North Cascades (Fig. 41a and b) is another such example.

Changes of a different nature in appearance of a glacier are illustrated in the three photographs of Figure 42. A small glacier on the upper flank of the Aiguille d'Argentière in the French Alps presented a thick layer of ice with a steep frontal cliff in 1899. By 1942 this thick layer and its cliff had entirely disappeared, presumably removed by a change in mass balance which favored ablation more than accumulation. After the next sixteen years not only had this glacier recession ended, but the

39. *Rapid glacier advance (surge), Tyeen Glacier, north branch, Fairweather Range*

40a. Build-up and discharge
of small, steep glacier, South
Mowich Glacier, Mount
Rainier, Cascade Range, 1960

40b. South Mowitch Glacier,
Mount Rainier, 1962

41a. Build-up and discharge of small, steep glacier, "Spillway Glacier," North Cascades, 1959

41b. "Spillway Glacier," 1960

1899

1942

1958

*42. Aiguille d'Argentière, French Alps,
in 1899, 1942, and 1958*

43. *Advancing glaciers, Roosevelt and Coleman Glaciers, Mount Baker, North Cascades*

44. Advancing terminus, Coleman Glacier, Mount Baker, North Cascades

formation of thick ice cliffs had far exceeded that in 1899. The exact combination of factors affecting these changes is difficult to ascertain, for a glacier like this one hanging on a steep mountainside is strongly influenced by such agents of accumulation and ablation as wind-drifted snow (which depends on storm directions) and both snow and ice avalanching. While the main part of the little glacier almost disappeared by 1942, the rocks below it, prominently visible in the 1899 photograph, are completely blanketed by a smooth layer of snow and ice in the later photograph. Apparently the agents determining the mass balance shifted in a fashion that favored more accumulation on the lower slope and less above. By 1958 a general net accumulation regime had been well established. A prominent bergschrund runs along the foot of the cliffs and steep snow slopes in all three photographs, rising above the active ice of another small glacier mass at the right side of the scene. This bergschrund has remained remarkably stable over a period of fifty-nine years, in marked contrast to the behavior of the glacier.

Advancing glaciers are in the minority today, although they were far more common in the recent past when a period of cool, wet climate culminated in the "neoglacial" advances of the seventeenth and eighteenth centuries. But even now local climate variations are causing a number of glaciers to advance. Glaciers in the North Cascades of Washington State, after many years of retreat, experienced a period of glacier rejuvenation in the decade beginning in 1945. Glacier advance, shown in Figure 43, was first observed in the Coleman Glacier on the north side of Mount Baker. Later its companion glacier, the Roosevelt, fed from the same accumulation area, also began to advance. Advance of the Coleman Glacier has continued, although the rejuvenation ended for most North Cascades glaciers during the exceptionally long and hot summer of 1958. Since the picture shown in Figure 43 was taken, the Coleman Glacier (in the center) has continued to advance across the stream coming from the Roosevelt Glacier. Figure 44 shows the Coleman Glacier terminus from the ground, the advancing ice a striking contrast to stagnant, retreating termini like the one in Figure 32. Shear planes break the ice wall as the glacier front is thrust forward, while debris picked up by the ice is released by melt at the exposed face and slides down, constantly reforming small moraines of rubble at the ice edge.

Moraines

Moraines have already been mentioned during the discussion of glacier advance and retreat. These debris accumulations are the product of glacier erosion and flow. Their occurrence in a landscape is prominent evidence that glaciers have been present at some time in the past. Moraines associated with present-day glaciers give a glimpse of the way in which similar features were formed on a much greater scale during the Ice Ages.

Lateral moraines are sharp-crested ridges of rubble that mark a valley glacier's flanks in its ablation zone. In this age of generally receding glaciers, they are left standing to mark a glacier's former extent (Fig. 45); between them is found the *ground moraine,* the mantle of debris deposited on the glacier bed. If glacier retreat is interrupted by temporary pauses, or if slight readvances occur, then other sets of lateral moraines, like those displayed in Figure 46, may be built parallel to the first.

Ablation moraine sometimes blankets the lower reaches of a glacier. It is produced as melt concentrates at the surface debris that has been gathered and incorporated into the ice during its journey. If this rock blanket becomes thick enough, it insulates the ice and thus severely inhibits further ablation. The Red Glacier in Figure 47 has developed such a thick surface moraine that extensive areas of vegetation have taken root, a development possible only on stagnant glaciers with little movement.

One of the earth's most fascinating landscape patterns is the striped appearance which *medial moraines* lend to large glaciers like those shown in Figures 48 and 49. These long, narrow ribbons of rock originate where two ice streams join together, carrying the debris eroded from the glacier margins or deposited there by rockfall into a common line that demarcates the contribution of the two tributaries. If several glaciers join to form a large valley ice stream, each may contribute a medial moraine at the juncture. Volume of ice flow from each branch is shown by spacing of the medial moraines; note the small contributed flow and closely spaced moraines in the center of Figure 49.

45. Lateral moraines and ground moraine, Easton Glacier, Mount Baker, North Cascades

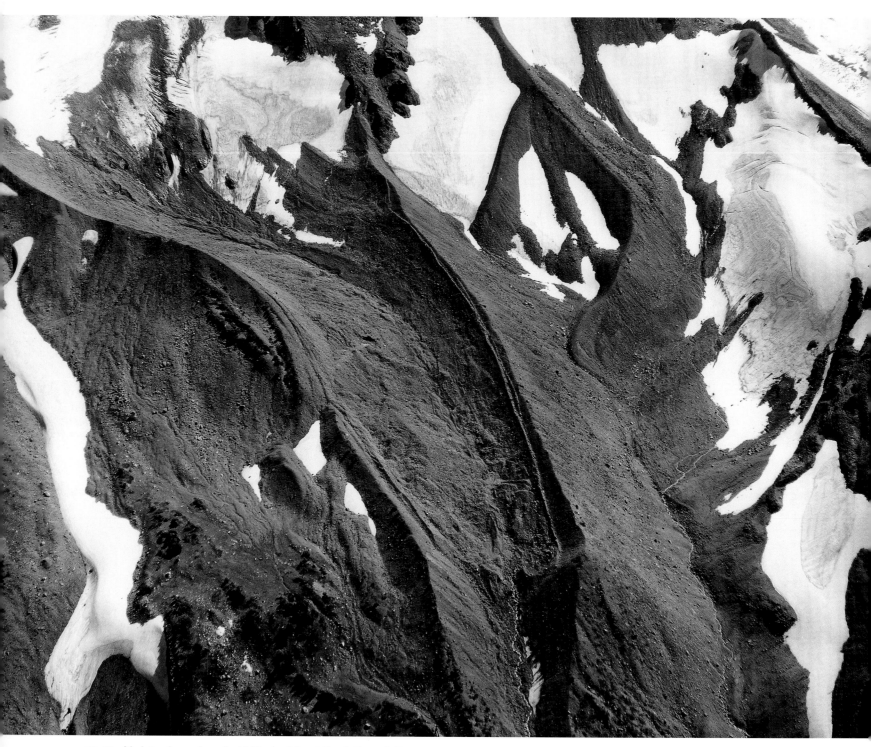

46. *Double lateral moraines, Ladd Glacier, Mount Hood, Cascade Range*

47. *(opposite) Surface moraine, Red Glacier, Mount Iliamna, Chigmit Mountains*

49. Medial moraines, "Speel Glacier," southeast Alaska, Coast Mountains

48. (opposite) Medial moraines, Kaskawulsh Glacier, St. Elias Mountains

Even among the moraine-rich glaciers of Alaska, the Yentna Glacier is distinctive. Its medial moraines originate from different geological structures, each with its own peculiar rock color. The result is a set of parallel medial moraines of different colors side by side, a difference so striking it is obvious even in a black-and-white photograph such as Figure 50.

Figure 51 shows the Malaspina, a huge (2,400-square-kilometer) piedmont glacier built by discharge of the Seward Glacier and various smaller valley glaciers onto the coastal plain of south central Alaska. A complex flow pattern, indicat-

ing periodic oscillation of ice discharge from the Seward Glacier, has built an intricate pattern of folded medial moraines as this sheet of ice spreads out across the plain. Medial moraines extend deep below the surface, hence more debris is concentrated at the surface as the ice is removed by melt. This explains why the moraine pattern, faint at the center of the Malaspina Glacier, gets stronger as the ice flows toward the edge. No one aerial photo presently shows the whole extent of this glacier. The sketch map of the moraine pattern shown in Figure 52 was pieced together from several dozen air photos.

51. *Distorted piedmont moraines, Malaspina Glacier, St. Elias Mountains*

50. *(opposite) Varicolored medial moraines, Yentna Glacier, Alaska Range*

52. *Moraine patterns of the Malaspina Glacier*

Glacier Surges

The Yanert Glacier (Figs. 53 and 54) offers a different picture of medial moraines. Twisted and looped into sworls, these moraines are more reminiscent of a chocolate marble cake than of orderly highways (Figs. 48, 49, and 50). Such contorted moraines, peculiar to a few large glaciers, are the product of glacier surges. The steep face of the terminus and extensive crevassing show that the Yanert was vigorously advancing in 1942. The advance of a surging glacier occurs on a different scale from that of a glacier which slowly grows in response to a climate change. The flow speed may reach several kilometers per year for short periods. Many unusual features of glacier surges have been identified by extensive aerial reconnaissance, and the physical characteristics of over two hundred surging glaciers have been observed in Alaska and western Canada. Recent research has pointed to the role played by accumulating meltwater at the glacier-bedrock interface in generating surges in glaciers.

After a long interval (fifteen to one hundred years) when the lower part of a glacier is relatively stagnant, a wave of ice from a *reservoir* area, located at some point in the glacier, abruptly begins to move down valley into a *receiving* area. The surface of the glacier is chaotically broken by the progressing wave, and ice may be displaced as much as 10 percent of the glacier length in a single year. The total glacier volume is not

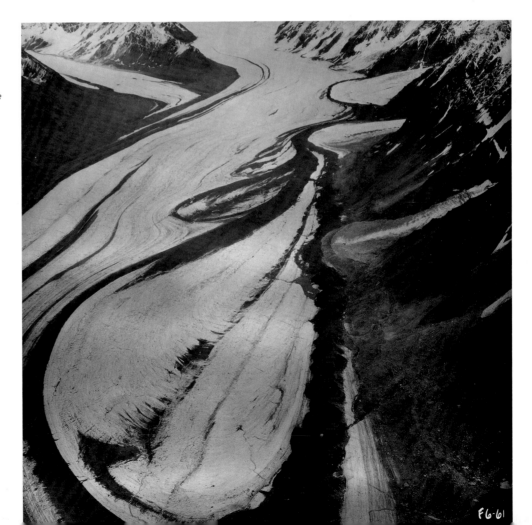

53. (opposite) Looped moraines and advancing glacier, Yanert Glacier, Alaska Range

54. (right) Moraine loops caused by successive surges of the main ice stream, Yanert Glacier, Alaska Range

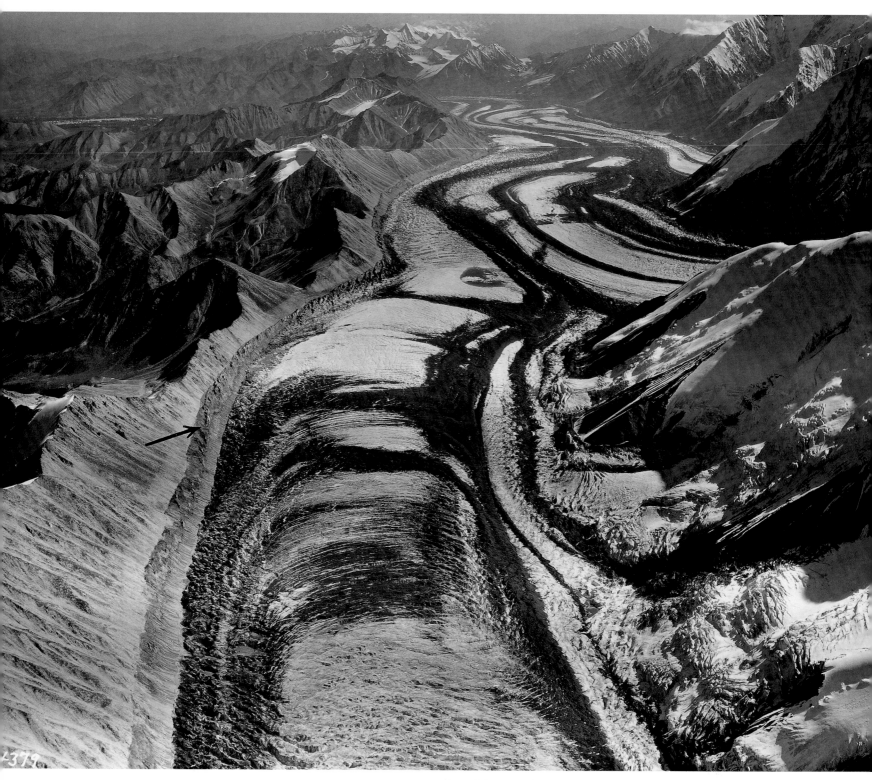

55. *Stranded marginal ice following glacial surge, Muldrow Glacier, Alaska Range*

augmented; ice for the advancing terminus is furnished by a lowering of surface level farther up the glacier. The stagnant ice in the lower part is activated, thickened, and pushed ahead by the wave. Only in exceptional cases does the glacier advance beyond its former maximum limit.

Surge Cycling. The active period of a surge generally does not exceed three years regardless of the size of the glacier. Surges occur repeatedly in the same glacier, although they are not always of the same magnitude. There is evidence that the largest displacements of ice occur at long, periodic intervals. Conterminous glaciers may not surge at the same time, and their frequency of surging may differ. There is little relation between glacier size and surge activity; small glaciers only a few kilometers long as well as large ones over one hundred kilometers long may exhibit this phenomenon. On the other hand, surging glaciers are found only in well-defined areas. This limited distribution strongly suggests that some external factor such as unusual bedrock roughness, permeability, or temperature is associated with the ability of a glacier to surge.

The surge of the Muldrow Glacier during the winter of 1956–57 is one of the best documented. Recognizable surface features were displaced downvalley as much as 6 kilometers, while the surface ice level in the upper part of the valley dropped 60 meters. Figures 55 and 56 show the stranded ice margins (arrow) left by this lowering of the glacier surface. The net ice volume transferred by this surge was about 3.7 cubic kilometers.

56. Close-up of stranded ice following surge, Muldrow Glacier, Alaska Range

The Black Rapids Glacier (Fig. 57) surged ahead 5 kilometers in three months during the winter of 1936–37. The looped medial moraines, similar to those of the Yanert Glacier (Fig. 53), result from a steady discharge from a tributary into a main ice stream which alternates between surge and stagnation.

The Walsh Glacier is another large valley glacier in the course of a surge at the time Figure 58 was taken in 1963. Note the looped moraines and extremely broken surface, both characteristic of surging glaciers.

57. *Looped medial moraine resulting from periodic surges, Black Rapids Glacier, Alaska Range*

58. Large surging glacier, Walsh Glacier, St. Elias Mountains

The most famous set of contorted medial moraines in Alaska is that of the Susitna Glacier (Fig. 59). The 1941 photograph by Bradford Washburn (Fig. 59a) has been widely published. Figure 59b, taken twenty-five years later, shows the moraine pattern completely changed. In 1941 the left-hand tributary was providing the major part of the ice flow, almost pinching off the stagnant main glacier. In 1952 a surge of the latter displaced the ice 4 kilometers downvalley. Since the surge, a new loop in the moraine has started to form as the tributary pushes into the again stagnant main stream. These changes can be more clearly seen in the sketch maps of Figure 60.

59a. The generation of looped and folded moraine patterns by periodic surges of a valley glacier with steady-state tributary, Susitna Glacier, Alaska Range, 1941

59b. (opposite) Susitna Glacier, 1966

SUSITNA GLACIER
showing ice displacements
resulting from 1952 surge

after surge

before surge

l 0 l 2 3 4 5 km.

60. *Sketch maps of Susitna Glacier surge and moraine patterns*

Ogives

Some of the most striking patterns in nature are presented by the regularly spaced arcs that appear on certain glaciers. These *ogives* occur as two types: wave ogives, or undulations in the glacier surface; and band ogives, or alternate dark and light bands on a smooth ice surface. Figures 61 and 62 illustrate these two basic types, respectively. Figures 63, 64, and 65 are other well-known examples of band ogives.

Ogives originate when a glacier flows down a steep, narrow icefall. Not all icefalls produce ogives, and very few produce the very pronounced ogives illustrated here. Several theories have been advanced to explain how these features form, but no one theory is yet general enough to account for all the observed varieties. It is known that glacier flow velocity is much higher in an icefall than in the rest of the glacier. One plausible theory suggests that this local acceleration stretches out the ice and exposes more area per unit volume as it passes through the icefall. The exposed area is further increased by crevassing and fracturing. The ice in transit during the summer melt season thus suffers more than its normal share of ablation, while that passing over the icefall in winter is protected by a

61. Wave ogives, Yentna Glacier, Alaska Range

62. *Band ogives, Mer de Glace, French Alps*

63. *(opposite) Close-spaced band ogives, unnamed glacier at head of Patmore Creek, British Columbia, Coast Mountains*

64. *Band ogives, East Twin Glacier, southeast Alaska, Coast Mountains*

65. *(opposite) Structured band ogives, Gilkey Glacier, southeast Alaska, Coast Mountains*

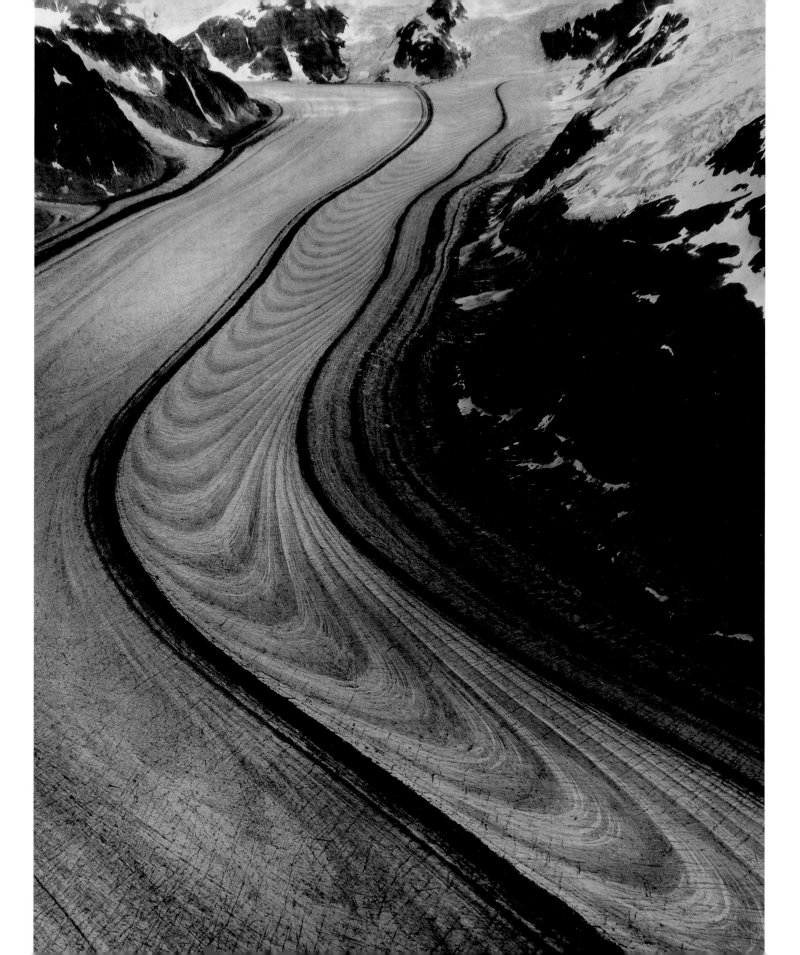

mantle of winter snow. The net result is annual fluctuation in the thickness of ice emerging at the bottom of the icefall even when flow velocity is constant with time. Waves develop further as these variations in thickness are compressed in the slow-moving ice below an icefall, as shown in Figure 66. In some cases an annual variation in flow velocity may add to the process as well. The alternate dark and light ogive bands are thought to originate in a similar fashion, the dark bands representing ice that has passed through the icefall in summer and accumulated extra surface dirt which is later concentrated by ablation. The Gilkey Glacier (Fig. 65) has an especially clear set of band ogives with fine structure visible in the bands. As

in the case of many band ogives, these result from wave ogives formed at the foot of the generating icefall, in this case that of the tributary Vaughn Lewis Glacier (Fig. 66).

It is far from clear why some ogive systems such as those in Figures 67 and 68 persist as waves a long distance down-glacier, while others (Figs. 63 and 64) quickly degenerate from waves into bands. The whole problem of ogive formation is far from solved, and future investigations may well lead to views different from those outlined above. It should be noted, for instance, that wave ogives are actually a rather superficial feature of the glacier ice; their amplitude is only a very small percentage of the total ice depth at the foot of an active icefall.

66. *Generating icefall for Gilkey Glacier ogives, Vaughn Lewis Glacier, southeast Alaska, Coast Mountains*

67. *Persistent wave ogives, Trimble Glacier North Branch, Alaska Range*

68. *Wave ogives buried under firn, Grand Pacific Glacier, Fairweather Range*

Why do some icefalls generate ogives while others in similar circumstances do not? The answer to this question is no doubt closely connected with some yet unresolved character of glacier flow.

The Gerstle Glacier ogive system is shown in Figure 69. That ogives have annual periodicity can be clearly demon-strated from Figure 70, with two photographs of the Gerstle Glacier icefall taken almost exactly a year apart. Debris features on the ice permit the same ogive waves to be identified in both pictures, and these have been numbered. In 1960 the crest of wave No. 6 was just forming at the icefall foot. A year later the ogive system had moved downglacier approximately one wave

69. Wave ogives and generating icefall, Gerstle Glacier, Alaska Range

70a. Close-up of Gerstle Glacier ogives, 1960

length in respect to adjacent landmarks; wave No. 6 had formed completely, followed by a trough, while the crest of No. 7 was just starting to form.

Ogives are a product of glacier flow dynamics and should be distinguished from the banding or layering that results from stratification of annual accumulation such as displayed in Figure 9. There was considerable confusion on this point among early investigators in the Alps, but the Scottish geologist J. D. Forbes correctly pointed out the difference as early as 1843, and band ogives are sometimes called "Forbes Bands" to this day. These two different patterns occur side by side on a single glacier in Figure 71. The faint ogives have a characteristic arcuate pattern, while the irregular bands delineate the uneven intersection of an ablation surface with the accumulation strata.

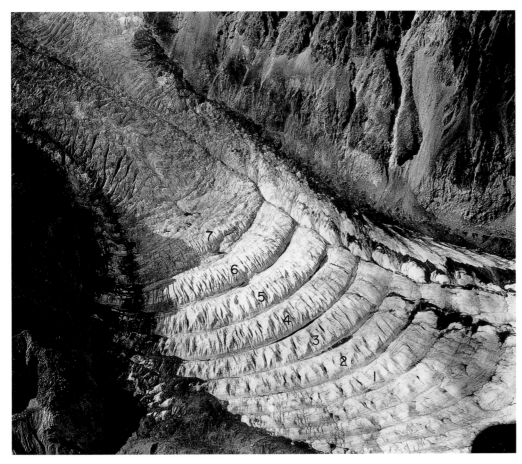

70b. Close-up of Gerstle Glacier ogives, 1961

In contrast to uniform, closely spaced normal ogives are the remarkable "false ogives" on the Margerie Glacier in the Fairweather Range, southeastern Alaska (Fig. 72). These do not originate from an icefall, and their wide spacing makes an annual periodicity implausible. In fact, faint annual ogives can be detected between the second and third heavy bands from the bottom of the figure. The Fairweathers present some of the steepest and most inhospitable terrain in North America, and large rock avalanches are frequent. Careful examination of photographs taken in different years shows that the dark bands on the Margerie Glacier represent debris from periodic rock avalanches which descended onto the glacier every ten years or so from the peak inside the curve where the glacier disappears from view in Figure 72. It is remarkable that such similar rock slides have occurred at such uniform intervals.

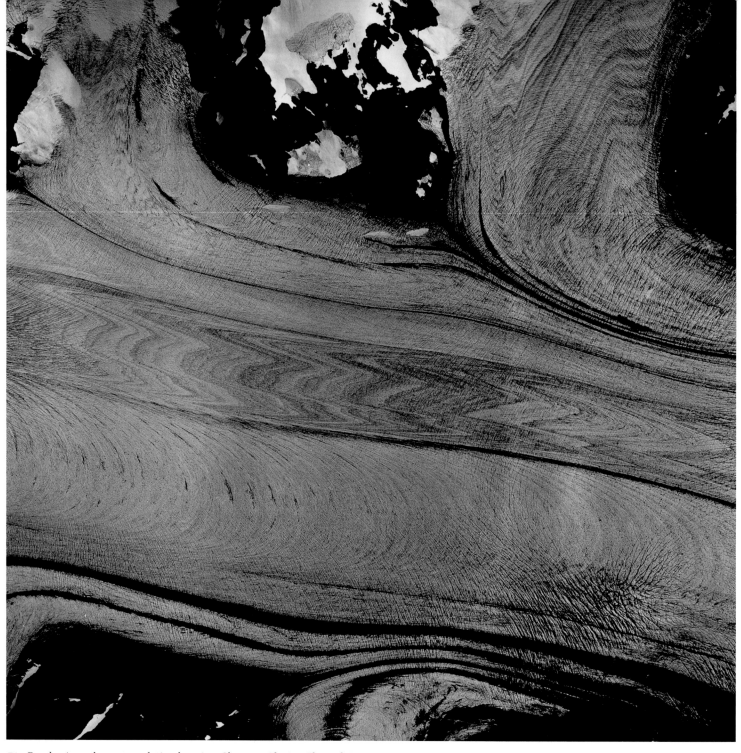

71. *Band ogives plus accumulation layering, Sherman Glacier, Chugach Mountains*

72. *(opposite) False ogives (debris from rock avalanches), Margerie Glacier, Fairweather Range*

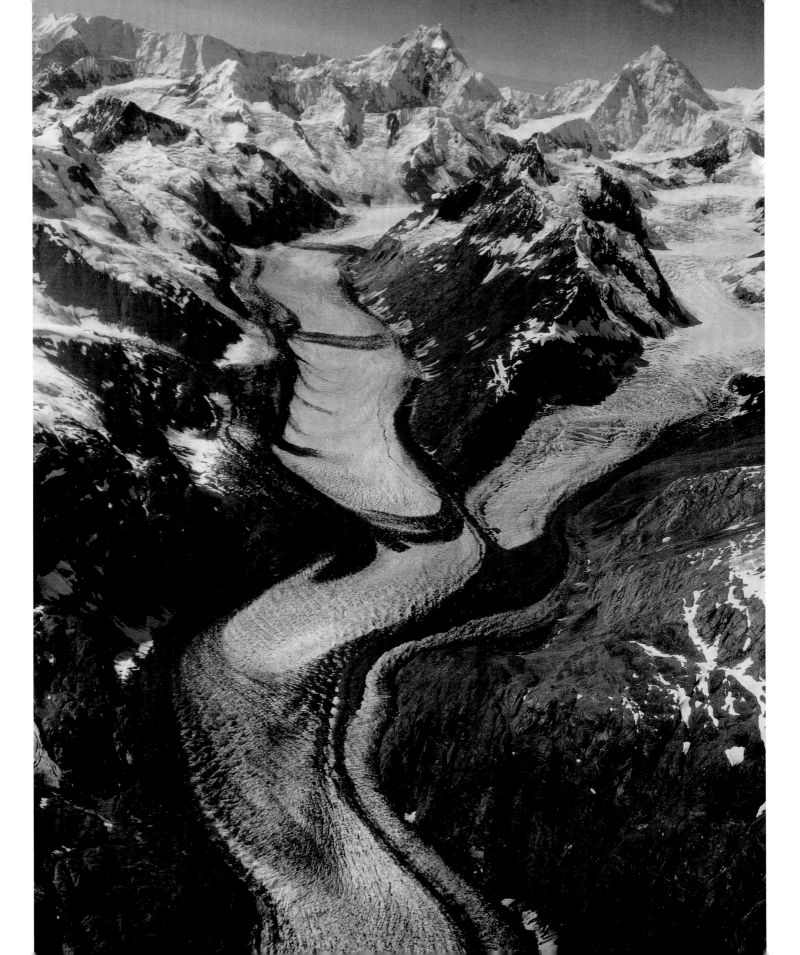

Meltwater on Glaciers

Glaciers everywhere except in the high polar regions experience a substantial amount of melt on at least part of their surface each summer. Meltwater like that shown in Figure 73 makes its way over, through, and beneath the glacier, eventually appearing at the terminus as a stream. The discharge of this stream fluctuates widely during the year. It also fluctuates widely during each day as the amount of meltwater released varies with daytime melt and nighttime freezing. The pattern of discharge from a glacier-fed stream can be just the opposite of that from a river draining a normal, ice-free valley. The latter has its maximum flow during winter months when rainfall is high, or during the spring snow-melt period. It is lowest late in the summer in regions where summer rainfall is slight. A glacier stream, on the other hand, discharges little water in winter because its source, the melt of ice, is cut off by freezing temperatures and by an insulating blanket of winter snow. Such a stream flows highest in the summer when glacier melt is at its peak. The presence of even small glaciers in a river basin thus tends to stabilize the total river flow, furnishing a maximum discharge of water just when other sources are at their lowest.

Streams often run for long distances over the surface of a glacier before finding their way down to its bed or off its margins. These rivers may develop the same dendritic patterns of tributaries (Fig. 74) as normal rivers running on land. On large glaciers these surface streams sometimes reach a very large size in summer, cutting deep channels with steep and dangerously slippery walls of ice. Few such streams follow the surface of the ice all the way to the terminus. Most eventually flow into a

73. Meltwater stream on glacier surface, Blue Glacier, Olympic Mountains

74. *Dendritic stream pattern on glacier surface, Black Rapids Glacier, Alaska Range*

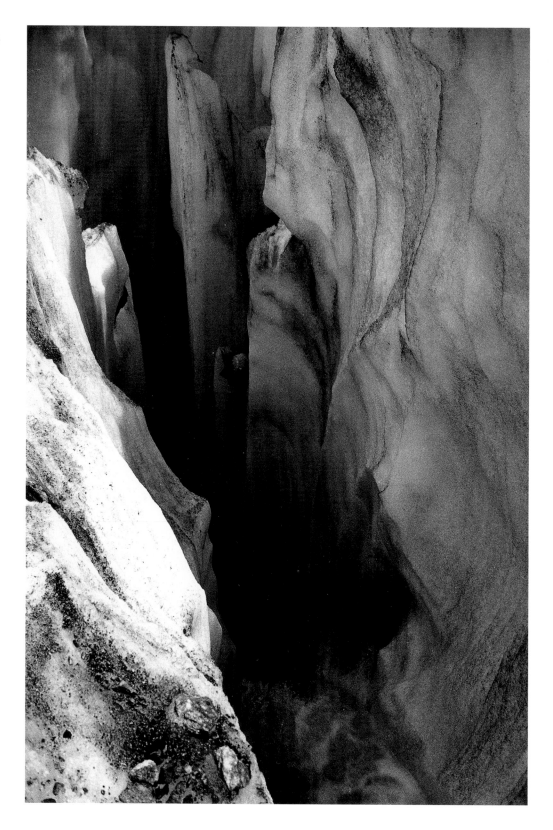

75. *Moulin, Glacier du Miage, Mont Blanc, French Alps*

subglacial stream channel which leads to the glacier bed by a sometimes intricate passageway. The hole where the meltwater stream plunges into the glacier depths is called a *moulin,* or *glacier mill,* a name stemming from the rumbling sound of the water, like that of an old-fashioned water-powered mill. Figure 75 shows a moulin with a vigorous stream plunging in at the bottom of the picture.

On the exposed blue ice of glacier ablation zones, meltwater flows on the surface because of the impervious character of the ice. In the accumulation zone, where the glacier surface layers consist of porous firn, the meltwater does not stand in open pools or streams at the surface except where it is prevented from draining by an ice layer. Sometimes a water table exists only a short distance below the surface, but is not visible. The apparently smooth snow surface conceals slush below. This condition can generate unpleasant surprises for the unwary glacier traveler. In Figure 76, a tracked vehicle has broken into one of these *snow swamps.* If the snow becomes sufficiently saturated on a gentle but perceptible slope, it can break loose in a "slush avalanche" which sweeps down the glacier at high speed.

76. Snow swamp, Greenland Ice Sheet

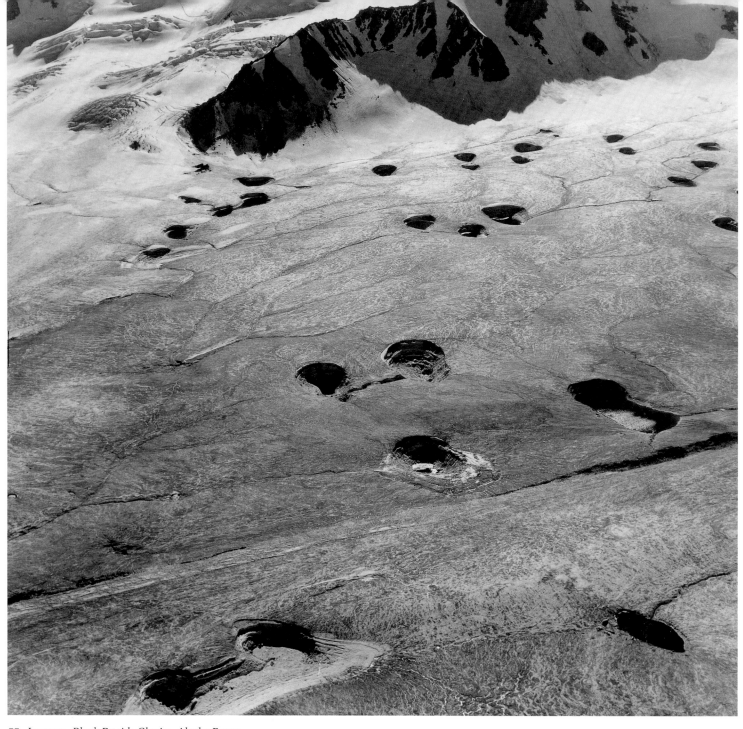

77. Lacunas, Black Rapids Glacier, Alaska Range

The dark pockmarks, known as *lacunas*, shown in Figures 77 and 78, probably owe their origin to perennial meltwater ponds on the surface of nearly stagnant, crevasse-free ice. Features on a much larger scale, called *ice dolines*, have been observed in Antarctica. Their formation has been postulated to be caused by depressions in the ice which fill with summer meltwater, then freeze over. Later glacier motion opens a sub-glacial passage which drains away the water, and the ice "lid" then collapses.

78. *Lacuna Glacier, Alaska Range*

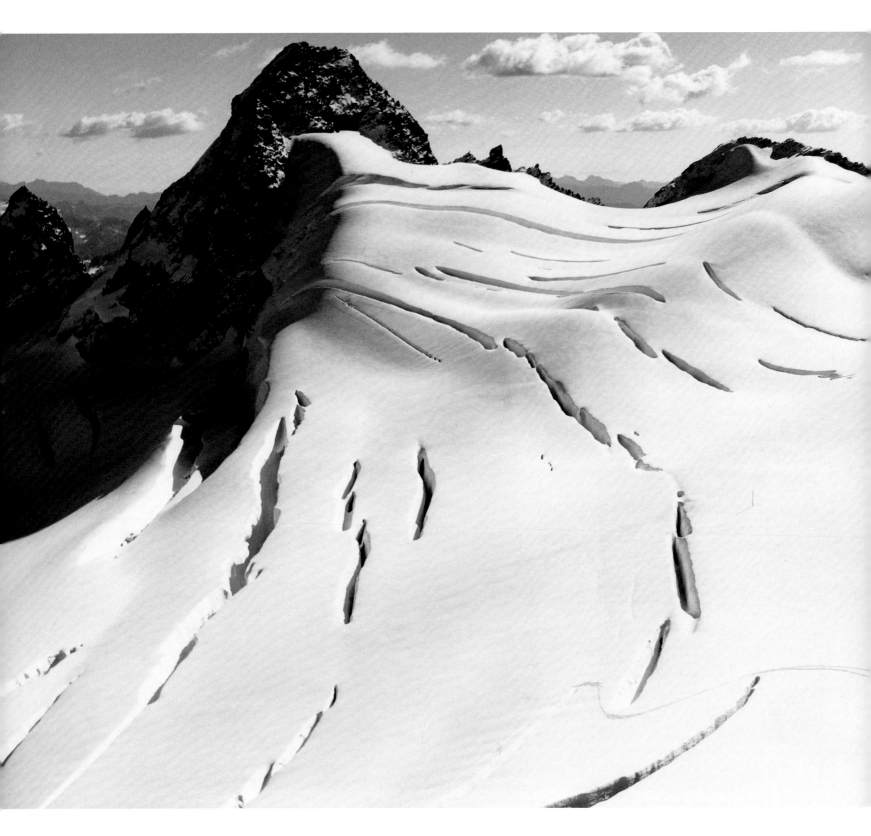

Surface Details

Most of the glacier features illustrated up to this point have been photographed from the air. This viewpoint is often required to examine the larger aspects of glacier ice, but there are many details of glacier ice that are better seen from the ground.

The most frequent glacier traveler is the mountain climber. Normally, shattered glacier surfaces will be avoided (Fig. 23), for ice broken to this extent is virtually impassable. An orderly arrangement of crevasses like that in Figure 79 is the sort of glacier terrain a mountaineer can traverse without difficulty. This picture shows an accumulation zone in a maritime climate of heavy snowfall. The crevasses are large and well developed, but spanned by numerous firn bridges and far enough apart to permit some freedom of movement. The accumulation zone of a glacier is a region of extending flow where the ice velocity increases downstream. This means that the surface is being stretched. Most of this stretch takes place by the widening of crevasses, to which is added the effects of melting. By late summer even a simple crevassed area like this may become nearly impassable through the collapse of snow bridges and extension of open crevasses clear across the glacier. Mountaineers then must thread their way along precarious routes like the one in Figure 80 (see also Fig. 19).

By all odds the most popular view of a crevasse is from above, looking in. The aerial photographer can view these pitfalls with impunity, and the wise mountaineer takes great care to avoid falling into glacial chasms like the one in Figure 81, a good example of a large ablation-zone crevasse that has been extensively eroded by melt. Those who do happen to find themselves plunged into such a crevasse (hopefully attached to a resilient climbing rope) rarely give serious thought to photography in their anxiety to regain the surface. Figure 82 is the view from the bottom of a crevasse, looking up. It was obtained under unusual circumstances. A recent scientific project on the Blue Glacier, located in Olympic National Park, involved digging a

80. *Ridge between firn crevasses, Blue Glacier, Olympic Mountains*

79. *Accumulation zone crevasses, Blue Glacier, Olympic Mountains*

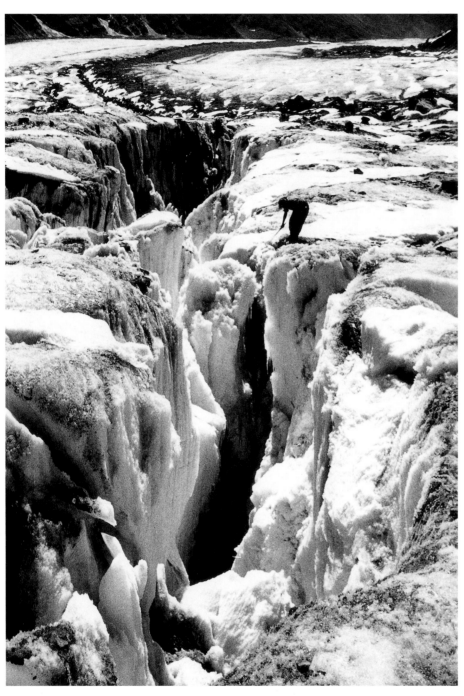

81. *Ablation zone crevasse, Chogo-Lungma Glacier, Karakoram Himalayas*

82. *(opposite) Inside of a crevasse, Blue Glacier, Olympic Mountains*

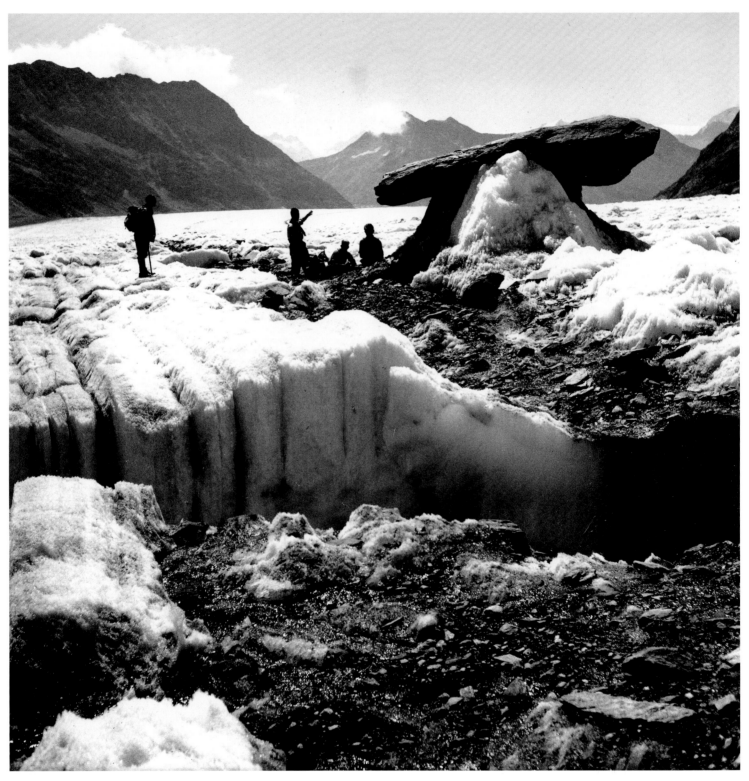

83. Glacier table and foliated ice, Great Aletsch Glacier, Switzerland

tunnel into an icefall to study sliding of the glacier over its bed. This tunnel intersected the bottom of a crevasse, permitting easy access (and egress) by which the photographer could obtain the picture at leisure. The intricate passages and rough surface are typical of ice that has been sheared and then exposed to melt, probably in this case by circulation of warm air from the surface.

The photograph in Figure 83, taken on the ablation zone of the Great Aletsch Glacier in Switzerland, illustrates a number of different glacier features. The rock debris mantling part of the ice is a medial moraine. Such rock material is distributed vertically through the glacier ice and becomes concentrated at the surface as ablation lowers the ice surface exposing the rocks. The dark area on the wall of the crevasse that traverses the foreground shows where the ice is filled with rock debris. The thin layer of dirt and rocks concentrated on the surface is darker than the surrounding ice, absorbs more heat from the sun, and contributes to faster ice melt. For this reason the debris-mantled surface is lower than that of clean ice. If the debris layer is thick enough, however, it insulates the ice from the extra heat and inhibits ice ablation. Ice so shielded stands above the level of the normally melting ice. If a rock is large enough to inhibit ice melt and shaped so that it does not readily tumble from the resulting pedestal, it becomes elevated above the surface as a *glacier table.*

On the left side of Figure 83, several vertical stripes on the wall of the crevasse are obvious extensions of similar stripes on the glacier surface. These stripes define the edges of planes of *foliation.* They are visible at the surface because these planes contain either slight concentrations of dirt, or ice whose capacity to absorb solar radiation is different from that of the surrounding ice. Either case leads to local alteration of melt and consequent irregularities in the glacier surface. Foliation is the layering in glacier ice produced by rapid deformation which causes the ice to recrystallize along certain planes. Hence it is a product of glacier flow that is completely unrelated to accumulation layering like that in Figures 10 and 11.

Not all the surface features on a glacier derive from water, rock, and ice. Surprisingly, living creatures make their permanent home on the glacier surface. The *ice worm* has been so widely celebrated in songs and stories of the northland that many regard it a myth, to be classed with the griffin and the unicorn. Contrary to popular supposition, it is not the product of imagination, it does not bear suspicious resemblance to a piece of cooked spaghetti with ink spots for eyes, nor is its

84. *Ice worms, Blue Glacier, Olympic Mountains*

normal habitat the bottom of a cocktail glass. The ice worm is a common inhabitant of the glaciers and permanent snowfields found in the temperate, maritime climate of the North Pacific Coast. It is, in fact, an annelid worm belonging to the class Oligochaeta and thus a relative of the common earthworm.

This peculiar worm thickly populates the glacier snow surfaces of early summer, and on a clear summer evening many hundreds to the square meter can be counted over large areas of a glacier. The worms in Figure 84 were not gathered together for a picture; they were photographed in their natural abundance on the summer snows of the Blue Glacier.

The ice worm probably has one of the most remarkable temperature environments of any creature on earth. Its entire life cycle is spent very close to the freezing point of water, subsisting on the pollen or algae incorporated in the snow by wind and concentrated at the surface by ablation. During the day the ice worm retreats 30 centimeters or more below the surface to avoid heat from the sun, but it comes out in the evening or during cloudy weather, presumably to search for food. As soon as radiation cooling on a clear night starts to freeze liquid water at the snow surface, the ice worm retreats once more, staying beneath the chilled nocturnal crust of frozen snow. A similar

85. *Sun cups, Taku Glacier, southeast Alaska, Coast Mountains*

cycle happens on an annual basis. Each autumn as the chill (subfreezing temperature) penetrates deeper and deeper into the firn, the ice worm retreats farther into the glacier. When spring melt dissipates the winter chill, it begins working up to the surface through all the accumulated snow, to appear in May or June. Such a life cycle can occur only on permanent snow and ice, and thus the ice worm is never found on transient winter snowfields. In fact, its presence or absence on a particular part of the glacier surface can clearly delineate the margin separating glacier ice from annual snowfields.

When snow or firn melts, the process of ablation never proceeds with complete uniformity. Slight irregularities in the snow structure, or local variations in melt, or both, soon produce a rough surface. There are several processes that can cause the roughness to increase as melt continues. Whenever this happens, the snow or firn surface soon develops a pattern of sun cups like those in Figure 85. These hollows in the snow

surface can range from shallow ripples only a centimeter or so deep to patterns with a relief of half a meter or more (Fig. 86).

Sun cups begin to form whenever weather conditions favor evaporation at the high points of the snow surface and melt in the hollows. There are definite conditions of temperature and humidity under which this is possible, with the additional requirement of an extra source of heat other than that from warm air. Normally this extra heat can come only from sunshine; hence the name "sun cups." About 7 1/2 times as much heat is required to evaporate a given mass of ice (or snow) as to melt it. Even if more heat reaches the high points of a rough snow surface than reaches the hollows, this may actually cause reduced ablation at the high points when enough evaporation takes place there. Thus the hollows ablate more rapidly, and the sun cups grow deeper.

Once the snow surface has been transformed from rough irregularities to regular sun cups, another mechanism begins which can continue even when favorable atmospheric conditions are not present. Percolating meltwater flows from the peaks down the sides of the cups to the lowest point in the center, then drains away through the snow. The finest particles of dirt concentrated at the surface by ablation are carried by this meltwater and congregate at the bottoms of the cups. The bottoms of these hollows grow darker than the crests, absorb more heat from the sun, and experience increased melt which further deepens the cups.

In higher latitudes where most glaciers are found, the sun seldom or never stands directly overhead. When not at the zenith, the sun shines more directly on one side of the sun cup than on the other, causing more melt on one side of the cup. This causes a field of sun cups to migrate as the cups deepen and form. In the Northern Hemisphere this migration is toward the north. Under favorable circumstances it can amount to several centimeters a day.

The most extreme weather conditions favoring sun cup development are found at very high altitudes in the tropics. Here the air is cold and usually very dry, the reduced atmospheric density augments evaporation, and intense heat is available from the sun. In such circumstances, sun cups keep on growing for weeks or months. The cups soon intersect, leaving parts of the ridges between standing as pinnacles. The greater the relief, the greater is the difference in ablation between pinnacle (evaporation) and hollow (melt). The ultimate result is a field of *nieve penitentes.* This name, Spanish for "penitent snow," originated in South America where the phenomenon is frequently seen in the high Andes. With the advent of warm weather, the tips of the snow pinnacles often droop or bend

86. Deep sun cups, Disappointment Cleaver, Mount Rainier, Cascade Range

over, giving the appearance of ranks of hooded monks with their heads bowed in prayer—hence the term "penitent." In Figure 87, the formation of hollows at the bases of the pinnacles has proceeded right through the winter snow cover to the ground, leaving in some places arrays of isolated pinnacles. The area in the foreground shows strong formation of nieve penitentes in snow; the background exhibits a similar formation in glacier ice. Figure 88 is an extreme example of nieve penitentes. Layering of the snow is plainly visible in the sides of the pinnacles.

High-altitude, low-latitude glaciers mantled with rock debris (ablation moraine) sometimes develop surface structures analogous to nieve penitentes. These ice pinnacles are built in part or sustained by the evaporation-melt processes described above, but they also experience a greatly enhanced effect from the difference in solar heat absorbed by the dark debris mantle and the white ice surfaces. Some of the ice pillars in Figure 89 might have started as supports for glacier tables (Fig. 83), but any random variation in debris cover of the glacier surface often serves the same purpose. Once clean ice stands above the debris, the same evaporation-melt process that enhances nieve penitentes can generate the pillars. Subsequent ablation then alters their shape to the triangular pinnacles of Figure 90 (see also foreground in Fig. 12). Under favorable circumstances these pinnacles may continue to grow to a very large size. They do not, of course, grow in the sense of getting higher at the top. The debris-covered ice melts downward much faster than clean, white pinnacles, so they are left behind as the remnants of a rapidly eroded glacier surface. The giant pinnacles like those in Figure 91 were named *ice ships* when they were first discovered on the Baltoro Glacier in the Karakoram Himalayas.

88. Highly developed nieve penitentes, Cerro Negro, Chile

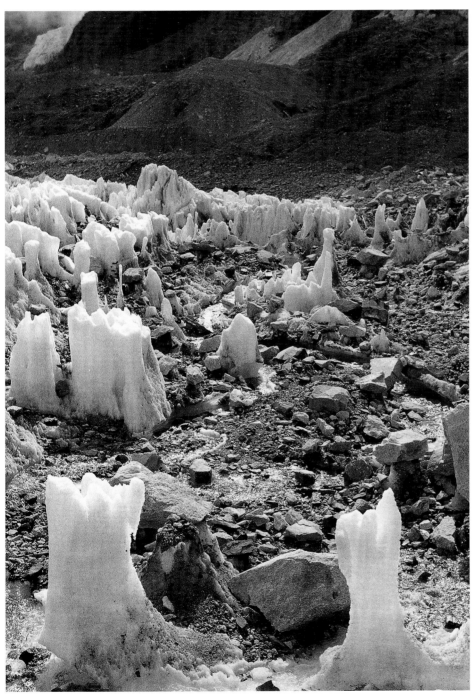

89. *Ice pillars, Khumbu Glacier, Mount Everest*

90. *Small ice ships, Khumbu Glacier,
Mount Everest*

91. *Large ice ships, Khumbu Glacier,
Mount Everest*

Calving Glaciers

Small glaciers that end on land advance or retreat by responding to climate effects within a few years or decades. Large glaciers also respond directly to climate variables but these responses are delayed by the time it takes for the changes in accumulation and ablation to affect the terminus. In addition to these effects, glaciers that end in water are subject to highly variable calving of *brash* and *icebergs*. Calving may be negligible in very shallow water, especially if the glacier is not very active, or can be gigantic as in the case of icebergs released from Antarctic ice shelves.

Passive calving. Slow flowing glaciers that terminate in fresh water lakes generally calve icebergs quite slowly regardless of water depth. Floating ice tongues may develop when such glaciers retreat. Icebergs large relative to the glacier can drift around the lake for years before melting. Quite similar conditions, except on a far greater scale, take place in large, subpolar glaciers such as Petermann Glacier (Fig. 125), or polar glaciers which terminate in ice shelves (Fig. 127), where icebergs sometimes as much as 100 kilometers in length are calved and drift off to sea. The term *passive calving* is appropriate across this wide range of sizes.

Active Calving. Quite another form of iceberg release takes place in moderate to rapidly flowing glaciers regardless of whether ending in fresh water lakes (Fig. 93) or salt water. Even glaciers which normally calve passively in proglacial lakes such as those at Bering Glacier, Alaska, calve actively during surges. Most of the spectacular *tidewater glaciers* that end in the sea calve actively. In Alaska, Greenland, and elsewhere, such glaciers often become tourist attractions. Since the glacier ice is badly fractured by rapid flow, the icebergs released are much smaller than those passively calved. Typically, these fall from high terminal cliffs or float up from subsurface releases in a spectacular fashion. Large waves frequently result, making such glaciers exceedingly hazardous to approach closely.

Calving Glacier Advance. As flowing ice constantly erodes its bed and adjacent mountain slopes, the resulting rock debris is transported downglacier and dumped at the terminus. In calving glaciers this forms a terminal moraine shoal inhibiting iceberg discharge. In vigorous, healthy glaciers with high accumulation area ratios (AAR), the ice is also eroding the back slope of the moraine and the debris is pushed over the crest of the shoal, and in this way the moraine is slowly moved ahead. This is a relatively slow process, one kilometer of advance per century is about as fast as most calving glaciers advance in Alaska, the speed dependent upon the glacier size and available moraine debris (Figs. 92, 97).

Extended Position. Once a glacier enters water and begins calving icebergs, a slow advance over several centuries or even thousands of years will take place by the above processes. In Greenland, Alaska, Norway, New Zealand, and South America, deep channels up to 50 or more kilometers in length have been eroded by such glaciers during the Pleistocene. These deep, steep-walled valleys produced by glacial erosion are called *fjords* when they are inundated by the sea. This most frequently results from a glacier eroding its bed far below sea level in these narrow, mountainous canyons. Fjords are typically much deeper than the coastal waters with which they merge. Many previously carved fjords are occupied in part by present-day tidewater glaciers, such as Meares Glacier, shown in Figure 96. When the ice retreats, it leaves behind a water-filled fjord like Tracy Arm (Fig. 95), which presents a remarkably accurate picture of what the Meares Glacier fjord would look like if the glacier were not present. Present-day glaciers readvanced down many of these fjords in the late Holocene (last 6,000 years). In many cases the fjords were almost completely filled by these advances to what is called an *extended position* (Fig. 94a).

Drastic Retreat. Most of the glaciers described above have retreated very rapidly in the past century or two, first by becoming over-extended due to continued slow advance, followed by climate change favoring glacier recession. Once a glacier recedes off the terminal moraine shoal into deep water, the calving rate increases up to an order of magnitude and annual ice loss due to calving far exceeds annual snow accumulation. It has been determined that the annual calving losses exceed accumulation on Columbia Glacier by about five times (Fig. 94b). Thus the glacier begins an irreversible *drastic retreat*.

Retracted Stable Position. Recession continues as long as the glacier ends in deep water regardless of climate effects

92a. 1961. Johns Hopkins Glacier (center) and Gilman Glacier (left), Glacier Bay, Alaska, exhibiting, respectively, substantial and little advance.

92b. 1986. In twenty-five years, Johns Hopkins Glacier advanced 1,100 meters due to moraine debris building a shoal. Gilman Glacier hardly advanced at all, owing to the absence of shoal-building.

which might otherwise favor stability or advance. Once the glacier reaches shallow water, however, the retreat usually ends and the terminus stabilizes.

Channel Configuration. Calving glaciers during drastic retreat are subject to yet other forms of instability. The *channel configuration*, that is, its length, width, and shape, as well as depth, is critical in determining the calving rate. Narrow channels regardless of depth impede iceberg discharge; crooked channels provide *pinning points* where the glacier maintains a tenuous stability. In such situations, the glacier retreat is delayed at each point, sometimes for several decades, then recedes catastrophically to the next obstruction (Fig. 96). Broad, deep, straight channels, on the other hand, allow almost unlimited calving, and the rapid recession proceeds unimpeded.

Calving Glacier Cycling. Even following drastic retreat, many calving glaciers continue to calve icebergs. In fact, due to the retreat the AAR may become remarkably high—as much as 0.96 being the case with Yahtse and 0.95 for the Hubbard Glaciers in Alaska. Thus, other than for calving losses, these glaciers are very healthy indeed and anything inhibiting calving will result in readvance. Which is just what happens as new moraine shoals develop. Such alternate slow advance, drastic retreat, and readvance is called *calving glacier cycling*. Because of the infinite variations in each glacier situation, advance and retreat cycling of even adjacent glaciers may be completely out of phase, such as the Columbia Glacier (Fig. 94) and Meares Glacier (Fig. 96). There is evidence of three or more such advance-retreat cycles for several Alaska glaciers, such as the Guyot and Hubbard Glaciers. The twenty-five-year readvance of Johns Hopkins Glacier, shown in Figure 92, follows a drastic retreat of this and many other glaciers filling Glacier Bay as late as 1750 AD.

93. Calving iceberg, Miles Glacier, Copper River Valley, Alaska

94a. *1961. Columbia Glacier, Alaska, discharging icebergs on a terminal moraine shoal. A large U-shaped embayment has formed in the 60-meter-high ice cliff, left. Increase in size of such embayments led to the glacier receding from the moraine in 1980, beginning a drastic retreat.*

94b. *(opposite) 1994. Columbia Glacier, showing the effects of fourteen years of drastic retreat. The glacier has receded 10 kilometers. Between the terminus (upper center) and the terminal moraine shoal (lower left), a tidewater basin as much as 360 meters deep and 3 kilometers wide has been exposed, which is jam-packed with floating brash ice and icebergs.*

95. Tracy Arm, southeast Alaska, Coast Mountains

96. *Meares Glacier, terminating in Unakwik Inlet, Chugach Mountains.*
This glacier has slowly advanced on a terminal moraine shoal more than
one kilometer since 1910.

Overleaf:
97. *Tidewater glacier and ice-choked fjord, Dawes Glacier, southeast*
Alaska, Coast Mountains

Glaciers, Volcanoes, and Jökulhlaups

The close association of glaciers and volcanoes produces a number of unusual glacier phenomena. Such association is known for the prominent volcanoes of the Pacific Northwest (Figs. 98–101), but probably is best documented for the glaciers and volcanoes of Iceland. Glaciers that blanket the upper flanks of steep volcanic cones produce an especially rapid erosion of the soft and friable layers of lava, pumice, and volcanic ash. Wind transport of dust derived from these materials quickly causes the accumulation of a dark surface layer which is concentrated by summer melt and enhances absorption of solar heat. Local acceleration of snow and ice melt takes place wherever fumaroles (steam vents) discharge steam and hot volcanic gases.

If a summit crater, or caldera, exists, it is an excellent catchment for snow. This may provide the zone of principal nourishment for several glaciers like those on Mount Rainier (Fig. 98). Prior to the 1980 eruption, the summit crater of Mount St. Helens generated a peculiar glacier which discharged through a breach in its eastern wall. The extreme narrowness of this ice stream, shown in Figure 100, earned it the descriptive name of Shoestring Glacier.

In 1980 Mount St. Helens provided a remarkable display of volcanic effects on glaciers. Increasing volcanic activity that spring led to minor eruptions which blanketed its glaciers with ash. Rising magma within the volcano caused the whole north side of the mountain to bulge outward. By May 17, pressure from this bulging split the northside glaciers lengthwise. The jagged fractures are indicated by arrows in Figure 101(a). The lower half of Figure 101(a) also exhibits tracks of mudflows generated as the intensifying geothermal heat melted the ice. The following day rising pressures finally set off a major eruption, which blew off the top of the mountain (Fig. 101[b]), essentially destroying the glaciers. With much of the higher-altitude accumulation zones of the former summit and crater now gone, little ice flow has been available to support glaciers on the flanks of Mount St. Helens. Glacier ice is now reforming within the quiescent new crater.

The hydrologic cycle of snow accumulation, ice flow, melt, and runoff is a regular one for most glaciers. The meltwater discharge follows a pattern of minimum flow in winter and maximum flow in mid-summer during the height of the melt season. In certain glaciers this pattern may be interrupted by the sudden, brief discharge of very large quantities of water. Such an abnormally large discharge may be irregular and take place at widely scattered intervals, or the occurrence may be annual. Enormous quantities of water released in a very short time by glaciers often cause catastrophic floods in the valley below. Close association of glaciers and volcanoes favors this type of flood in Iceland, where it is sufficiently common to have been given a name, *jökulhlaup*. This term is now used elsewhere to describe catastrophic glacial floods.

The jökulhlaup originates from large bodies of meltwater impounded by the glacier along its margin, beneath, or even within, the ice. The outlet for these lakes or caverns can be closed by ice flow when the reservoir is depleted. The water level, rising with advent of the ablation season, heavy rainstorms, or through melt caused by volcanic heat, fills the reservoir. The initial breakthrough starts with a trickle of water through, over, or under the ice. If the impounded water is sufficiently deep, the hydrostatic pressure of the water can also exert sufficient force on ice at the bottom to deform it by plastic flow. The "leak" may also start through natural cracks, or simply by the water overtopping the ice. Once the discharge starts, flow water even slightly warmer than 0°C rapidly melts and enlarges the channel. The larger the channel becomes, the more rapidly the water escapes, providing additional heat to melt more ice. Such a sequence of events leads to an exponential rise in water discharge; the result is a sudden "dumping" of the impounded water in a matter of hours.

The Tulsequah Glacier, one of the outlet glaciers of the Juneau Ice Field, is famous for its annual jökulhlaup, which originates from glacier-dammed Tulsequah Lake, located along its west margin. Figure 102 shows the lower reaches of this glacier, with the site of the lake indicated by the arrow. This lake occupies a deep valley into which flows a distributary arm of the glacier shown in Figure 103. Each winter and spring this valley fills with water to a depth of about a hundred meters, indicated by a clearly visible strand line. The breakout usually comes in July, when a subglacial passage is opened beneath the

98. Ice-blanketed volcanic cone, Mount Rainier, Cascade Range

104

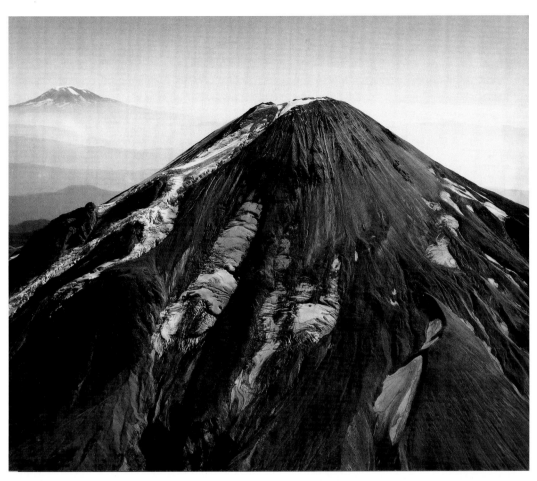

99. Volcanic cone with several small glaciers, Mount St. Helens, Cascade Range

100. Shoestring Glacier originating in volcanic summit crater, Mount St. Helens, Cascade Range, prior to 1980 eruption

glacier. This photograph of the lake was taken in 1961 about a month after the annual breakout; already the runoff from surrounding glaciers has begun to refill the basin.

Less clear is the origin of infrequent, erratic floods from glaciers mantling the steep flanks of volcanic cones. Here there is no obvious reservoir in which large quantities of water can be stored for sudden release, such as the glacier-dammed valley of Tulsequah Lake. Yet at infrequent intervals a destructive flood may emerge without warning from a glacier and sweep down the valley below. The glaciers of Mount Rainier have yielded a number of such floods in historical times, and geological evidence indicates that others have occurred in the past. The Kautz Creek Flood of 1947 was unusually destructive, for it destroyed over 1 kilometer of the glacier, cut a deep new stream

101a. Mount St. Helens, Cascade Range, May 17, 1980

101b. Mount St. Helens, May 18, 1980

103. Glacier-dammed lake, Tulsequah Lake and Glacier, British Columbia, Coast Mountains

102. (opposite) Tulsequah Glacier, producer of annual jökulhlaups, British Columbia, Coast Mountains

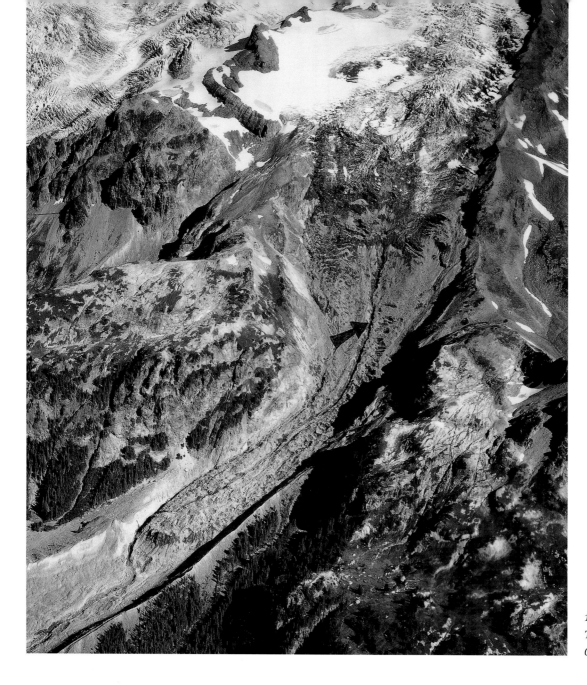

104. *Flood channel on South Tahoma Glacier, Mount Rainier, Cascade Range*

channel, deposited mud and rocks 10 meters deep over a high-way, and inundated a mature forest. Stratigraphy of the new stream banks disclosed that the Kautz Glacier had previously discharged similar floods. Though the 1947 flood followed an exceptionally heavy rainfall, a sudden release of water by the glacier was obviously involved. Other floods in this area (for example, the Kautz Glacier Flood of 1960) were not associated with rainfall. A more recent example is seen in Figure 104 on the South Tahoma Glacier. A hidden reservoir of water suddenly broke out in midglacier and flowed over the ice, eroding a deep channel (arrow) before the cavern drained and the flood sub-sided. These floods clearly resulted from the sudden discharge of water stored within or under the glaciers, and it is difficult to conceive how such large englacial or subglacial cavities can exist in an actively flowing ice. It has been suggested that heat suddenly released from within the dormant volcano may pro-duce the water by melting ice at the bed of the glacier.

Effects of Glaciers on the Landscape

Glaciers are very active agents of geological erosion; much of the spectacular mountain scenery of the earth has been sculpted by ice. Although ice itself is a very soft material with little abrading power on rock, glaciers carry with them much rock debris embedded in the ice, especially in the sole (see, for instance, Figs. 30 and 31). This rock debris serves as an effective tool with which the ice gouges and grinds the underlying bedrock during its journey down the mountainside. Bedrock exposed by a retreating glacier has imprinted on it the evidence of this erosion in the form of grooves and striations generated by rock particles of varying sizes in the glacier sole. Figure 105 shows such striations in freshly exposed bedrock along the margin of the Blue Glacier. Direction of ice motion was from upper right to lower left. The larger pattern of grooves or undulations is also visible, as well as an area just above the ice axe where a large piece of rock has been detached by the plucking action of the ice. Such plucking is one way the glacier acquires the rock tools for erosion.

One product of rock grinding against rock is *glacier flour,* the extremely fine dust that remains suspended in glacier-fed streams, giving them a milky appearance, and which even remains suspended in lakes fed by such streams. The content of rock powder gives these lakes unusual colors, ranging from muddy chocolate to brown, tan, cream, and even a brilliant emerald green.

The rapid erosion by glaciers deepens and broadens stream valleys into U-shaped canyons. These canyons and the mountain peaks resulting from ice erosion and severe frost action are steep-walled landforms very different from the smooth-crested, often intricately gullied slopes produced by other weathering agents. If composed of poorly consolidated rock, glacially steepened slopes are unstable and subjected to large-scale catastrophic collapse. Figure 106 shows an example of such a catastrophe. The Alaska Good Friday earthquake of 1964 dislodged many landslides among the glacier-carved peaks of the Chugach Mountains. One of the most spectacular was this one, which blanketed the surface of the Sherman Glacier with a huge mantle of rock debris. This slide originated from the face of Shattered Peak, just below the towering cumulonimbus

clouds in the center of the photograph. It flowed at extremely high velocity 6 kilometers over an intervening ridge and then at a very low gradient across the Sherman Glacier. It has been postulated that this remarkable flow was due to the formation of a lubricating air cushion under the rapidly moving debris.

The erosive action of moving ice causes a glacier to cut its bed deeper and deeper, broadening the valley floor and cutting back into the headwall of its accumulation basin to form a *cirque.* The character of the mountain terrain thus developed depends on the structure and quality of the bedrock as well as on the nature of ice flow. Massive, granitic bedrock especially tends to form spectacular mountain scenery under the sculpturing of glacier ice. An early stage of glacial erosion in granite is seen in Figure 107, where scattered cirques have been cut

105. *Glacier striations in bedrock, Blue Glacier, Olympic Mountains*

106. *Massive rock avalanche resulting from earthquake shaking of cliffs steepened by glacial erosion, Sherman Glacier, Chugach Mountains*

107. *(opposite) Biscuit-board topography, Wind River Mountains, Wyoming*

108. *Serrated arêtes, Eldorado Peak area, North Cascades*

into an elevated tableland, forming a *biscuit-board topography.* When glaciation is prolonged beyond the biscuit-board stage, the cirques and glacier-carved valleys eventually intersect. The jagged ridges that mark these lines of intersection are called *arêtes,* illustrated in Figure 108. On a larger scale, deep, glacier-carved gorges like the gigantic cirque in Figure 109 are also ringed with sharp arêtes. When ice has intensively eroded

109. *(opposite) Glacier gorge and arêtes, tributary to Johns Hopkins Glacier, Fairweather Range*

114

110. Arêtes and pinnacles in granite, Kichatna Mountains, Alaska Range

111. (opposite) Horn-type mountain, Devil's Thumb, Boundary Peak #71, Alaska, and British Columbia, Coast Mountains massive granite, the arêtes stand to the sky in dizzying pinnacles and ridges that attract the more daring mountaineers. The Kichatna Mountains in Figure 110 are a good example of this stage of glacial erosion. In regions of heavy snowfall and prolonged glaciation, ice has worn away all except the most durable and resistant rock outcrops, and horns are left as the only remnant of a long-vanished upland (Fig. 111).

The effects of glacial erosion (Figs. 107–111) are primarily those of scattered alpine glaciers, although in some cases, as seen in Figure 111, there is definite evidence that the terrain

112. *Juneau Ice Field, southeast Alaska, Coast Mountains*

was once inundated by a more extensive ice cover. *Ice sheets* presently cover practically all of Greenland and Antarctica. Comparable ice sheets covered areas in Asia, Europe, and North America during the Pleistocene. Several much smaller *ice fields* exist today. The Juneau Ice Field in southeast Alaska shown in Figure 112 is one of the best-known examples. Here a sea of ice fills the valleys, and only the highest ridges and peaks remain

uncovered. These areas of bedrock completely surrounded by ice are called *nunataks,* a word of Greenlandic Eskimo origin. The Bagley Ice Valley (Fig. 113) contains row after row of nunataks, each row consisting of the highest peaks of an ice-flooded mountain range.

When a landscape becomes almost completely buried under an ice field, the rough, irregular peaks and ridges are

113. *Ice field and nunataks, Bagley Ice Valley, south central Alaska*

worn away by the ice. Steep-walled valleys occupied by the principal glaciers draining the ice field are smoothed and rounded. When the ice vanishes, a more subdued landscape remains, broken by deep valleys and an occasional high, rocky promontory which as a nunatak escaped inundation. Figure 114 shows such a landscape.

Icecaps, generally defined as ice domes from which no

mountain peaks emerge, are most frequently found on plateaus or uplands in polar regions.

When glaciers melt, they uncover distinctive glacial deposits from which many features of ice erosion can be deduced. For instance, melting of a remnant of the Muir Glacier, on the east side of Glacier Bay, has exposed an unusual display of glacial geology (Fig. 115). The glacier had recently overridden

119

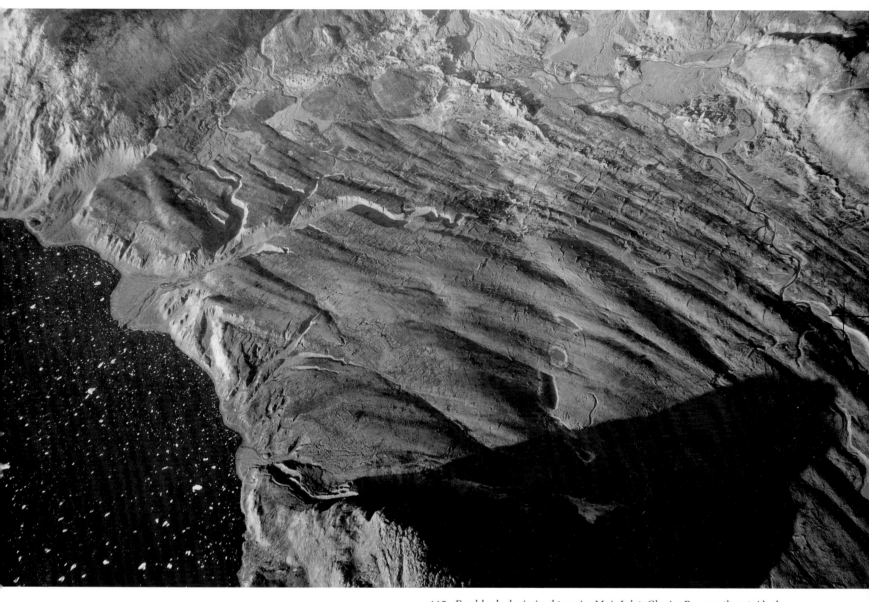

115. *Freshly deglacierized terrain, Muir Inlet, Glacier Bay, southeast Alaska*

114. *(opposite) Formerly ice-covered terrain, British Columbia, Coast Mountains*

this area of poorly consolidated sediments, and the flowing ice produced the series of gently rounded, parallel ridges aligned with the direction of flow. These are elongated forms of *drumlins*, or mounds of glacial-sculptured material left behind in an orderly array by a retreating glacier. Running across at right angles to these drumlin ridges are many narrow streaks of debris which stand out sharply, owing to their fresh and uneroded appearance. This debris, which had collected in the bottoms

121

116. *Glacial till, Jmjia Glacier, Garwhal Himalayas*

of crevasses, was lowered undisturbed onto the ground as the ice melted. Rapid incision of stream channels into the freshly exposed ground has exposed various former glacier outwash deposits, some containing forest remains, and layers of glacial *till* deposited by previous glacier advances. Glacial till, illustrated in Figure 116, consists of unstratified and largely unsorted rocks, gravel, and finer debris all jumbled together.

An *esker* is a ridge of sediment deposited by a subglacial meltwater stream. It follows the twists and convolutions of the generating stream. Eskers can form only in stagnant ice where the sediment can build up to appreciable depths and be left in place undisturbed when the ice melts. The twisting channel of an esker-building stream is especially well delineated in Figure 117. This esker, near the edge of the Malaspina Glacier (Fig. 51), has the additional peculiarity of being deposited on ice rather than on bedrock. Evidently it was formed by a meltwater stream whose channel did not extend to the glacier bottom. Two rough irregular eskers are also visible in the upper part of Figure 115.

Figure 118 shows an area of terminal moraines from which the generating glacier retreated a century or more ago. Many of the fine details of glacial deposits visible in Figure 115 were no doubt also present here, but have long since been obscured by erosion and concealed by thick vegetation. The outlines of some of the old terminal moraines can still be seen as patterns in the timber growth. Patterns like this are easy to recognize from the air but often difficult to detect on the ground.

117. *Esker, Malaspina Glacier, St. Elias Mountains, Alaska*

118. *(opposite) Forested moraines, Herbert Glacier, southeast Alaska, Coast Mountains*

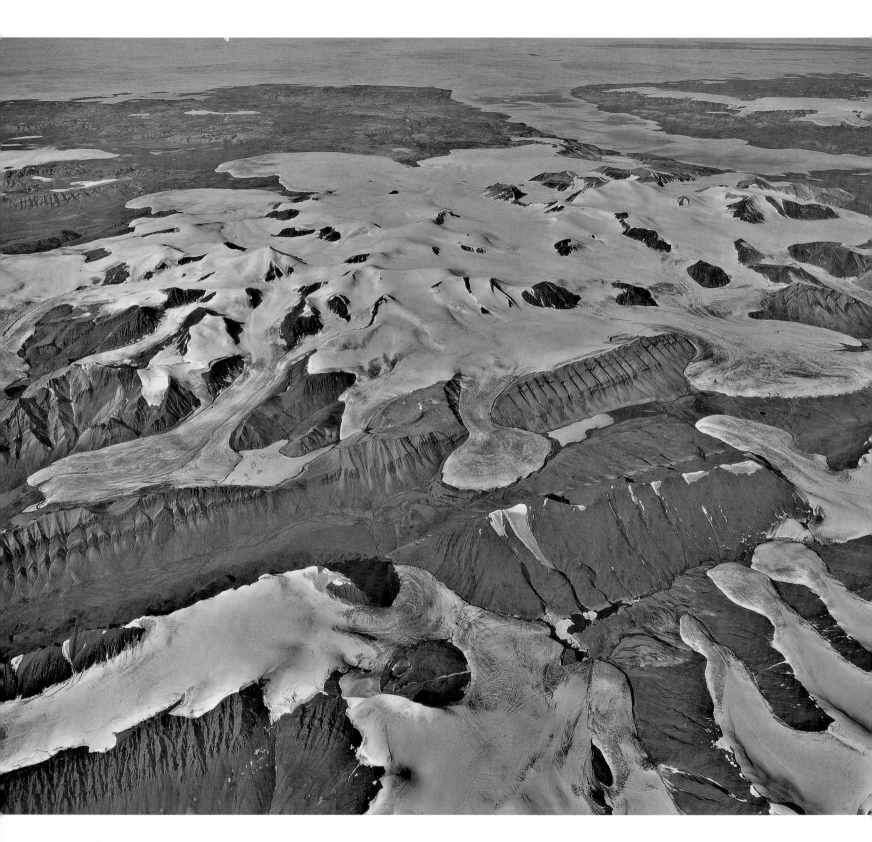

Temperate, Subpolar, and Polar Glaciers

There has been a common feature in the illustrations presented up to this point. All of the glaciers have been *temperate glaciers,* whose ice is very near the pressure melting point throughout except for a surface layer that is chilled below freezing each winter. The mode of flow of temperate glaciers is profoundly affected by the presence of liquid water throughout most of their mass. The ice deforms, fractures, and slides readily, and the regelation process is able to assist sliding of the ice over bedrock. The overall rate of glacier flow depends in part on the amount of liquid water produced by rainfall and surface melting.

Subpolar glaciers, on the other hand, exist perennially at subfreezing temperatures throughout most of their mass. Some liquid water may be present at the surface and in limited channels during the melt season. Glaciers in very cold climates may also be frozen in part to their bed, with the regelation process eliminated as a mechanism for sliding. The viscosity of ice rises with falling temperatures; coupled with a reduced activity index common to low temperatures and light precipitation, this leads to a substantially different mode of flow and a notably different external appearance of subpolar glaciers. The glaciers in Figure 119 obviously differ from those illustrated previously. The small ice field and surrounding individual glaciers behave much as if the ice were a layer of thick molasses poured over the landscape. Figure 120 is another example of such a phenomenon. The small esker along the margin of the glacier at the bottom of Figure 120 suggests that a substantial amount of meltwater must have flowed previously under the glacier. In cold climates, liquid water may be present in channels in ice that is below freezing temperatures for the most part.

The glaciers of Figures 121 and 122 are further examples of subpolar glaciers. These all serve as outlet glaciers stemming from extensive ice fields. Both figures illustrate another feature common to these glaciers—the steep, clifflike front of the terminus. There is also a noticeable absence of the prominent lateral

and terminal moraines that so often surround the lower ends of temperate glaciers. Both the steep ice fronts and the lack of moraines result from the relatively slow motion and the limited basal sliding. Such glaciers experience a mass balance substantially different from that of temperate glaciers. Precipitation is generally low and is nearly always in the form of snow. Ablation is also low, and meltwater runoff is sometimes almost absent. Where surface melting does occur, it usually is small in quantity and confined to the terminal region of the glacier. The activity index is low. Much of the ablation that does occur is along the periphery, where absorption of solar heat by the dark ground causes local melting and evaporation. This latter feature, combined with the relative immobility of ice, leads to the undermining and breaking off of ice at the glacier edges and consequent cliff formation. The same immobility greatly inhibits the glacier's role in erosion and internal transport of debris; hence the accumulation of morainal detritus is much reduced.

In true *polar glaciers,* subfreezing temperatures are present throughout their mass. Liquid water is restricted to very small amounts of rain or snow melt at the surface. After percolating below the surface, this water freezes immediately. Polar glaciers are frozen to their bed, and the regelation process is completely eliminated. They are found only in the high polar regions, such as parts of Antarctica and northern Greenland.

Although temperate glaciers have received the major emphasis in this book, they actually represent only a tiny portion of the world's glacier ice. Temperate glaciers are the best known and the most frequently visited because in such areas as the European Alps, Norway, and western America, they exist in the midst of populated areas. However, most glacier ice in the world is polar or subpolar. The Greenland and Antarctic ice sheets are the storehouses of some 99 percent of all existing glacier ice, and Antarctica alone contains around 70 to 75 percent of the world's fresh water locked up in ice. These ice sheets are continental in scale, burying whole mountain ranges, basins, and vast plains, and making alpine glaciers seem minor indeed in both areal extent and volume.

The weight of these ice sheets presses sections of the

119. *Small polar ice field, Axel Heiberg Island, Nunavut Territory*

120. *Subpolar glacier with piedmont tongue, unnamed glacier near Dobbin Bay, Ellesmere Island, Nunavut Territory*

121. *(opposite, top) Subpolar glaciers, Axel Heiberg Island, Nunavut Territory*

122. *(opposite, bottom) Subpolar glacier, "Polaris Glacier," Hall Land, northwest Greenland*

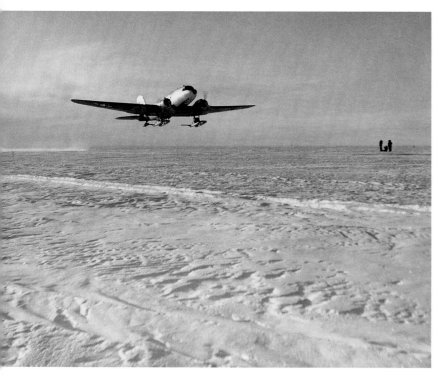

123. Surface of Greenland Ice Sheet

earth's crust down into the mantle so that the buried land beneath them has a lower elevation relative to sea level than it would if the ice were removed. Similar ice sheets once covered portions of North and South America, Europe, and Asia during the Pleistocene. Other parts of the world, such as North Africa, were ice covered in much earlier glacial periods.

The central plateau of a continental ice sheet, seen in Figure 123, presents the most monotonous scenery on earth. Mountain ranges are completely buried, so that no nunataks break the featureless plains of snow which stretch without perceptible slope or relief for thousands of kilometers. This accumulation of snow, century after century, has formed glacier ice thousands of meters thick. Ice sheets are so large that they make their own climate and affect the behavior of storms over large areas of the earth's surface. At the high altitudes and latitudes necessary to sustain ice sheets in today's climate, the persistent low temperatures (annual mean temperatures as low as -20° to -50°C) keep the air very dry, so that precipitation over Greenland and Antarctica in most places is very small, amounting to only a few centimeters of water equivalent each year. Even with such light accumulation, many cubic kilometers of snow are

added to the ice sheets each year simply because of their enormous extent.

The ice in these immense ice sheets slowly flows outward from the central plateaus and descends to sea level, where iceberg discharge eventually delivers it to the sea. Figure 124 shows an edge of the ice sheet in northwest Greenland, where streams of ice spill down from a high plateau into a fjord covered with sea ice. Such land is desolate and inhospitable. Much polar terrain like this is from the climatological standpoint a true desert because of the dearth of precipitation. Despite this, water in large quantities is present as perpetually frozen glacier ice and permafrost.

Figure 125 shows one of the least visited scenic wonders of the world, Petermann Fjord in northwest Greenland. This fjord is far larger in scale than those in Alaska illustrated earlier (for example, Fig. 97). It is 16 kilometers wide, and the vertical cliffs are 800 meters high. This photograph shows the boundary between Petermann outlet glacier, which stretches away into the distance, and sea ice in the foreground. The glacier occupying Petermann Fjord is afloat for a distance of over 100 kilometers, extending far beyond the horizon visible here.

In some places the outlet glaciers of ice sheets discharge large quantities of ice into the sea, where coastal terrain inhibits ablation by iceberg discharge. The glaciers then spread out over the sea surface to form an *ice shelf*. The small shelf in Figure 126 has formed from the outflow of an ice field on the southeastern side of Ellesmere Island in the Canadian arctic. This illustration shows the accumulation areas in the mountains, the ice field, and the ice shelf, all in a single view. Large sections of stagnant "fossil" ice shelves on the northern side of Ellesmere Island occasionally break off and are carried away by ocean currents. Trapped in the Arctic Ocean, they often drift around this basin as *ice islands* for years before they eventually escape to the south to break up and melt away.

124. (opposite) Subpolar glaciers descending from an icecap, Bessels Fjord, northwest Greenland

Overleaf:
125. Sea ice and Petermann Glacier, Petermann Fjord, northwest Greenland

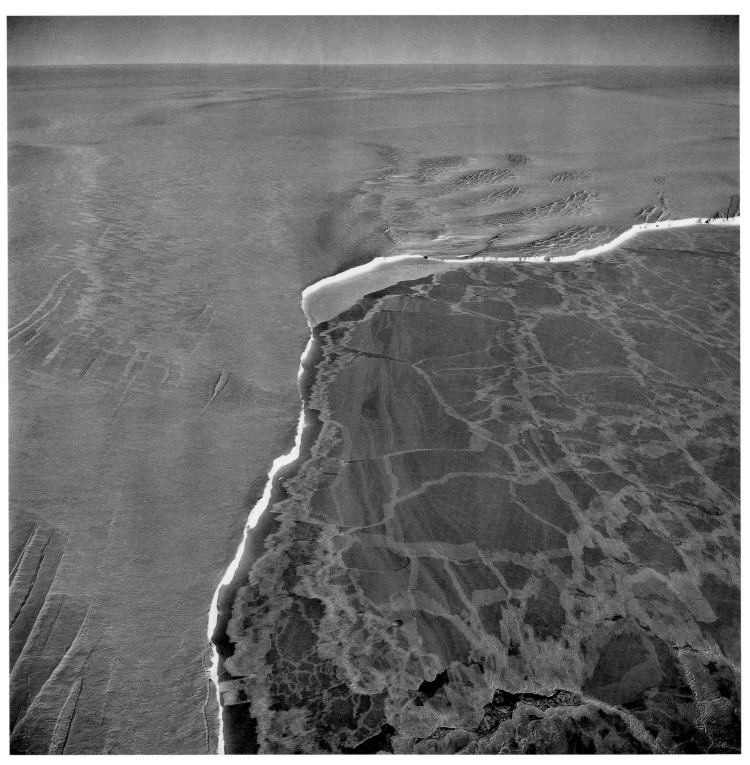

Overleaf:

127. *Floating ice shelf and barrier cliff, Ross Ice Shelf, Antarctica*

126. *Ice field and small ice shelf, Ellesmere Island, Nunavut Territory*

A single view of the major ice shelves is not possible in Antarctica, except from a satellite or space ship, for the scale is much too large. Figure 127 shows a section of the Ross Ice Shelf near Kainan Bay. The ice shelf stretches away into the distance, separated from the sea ice in the foreground by the bright line of its terminal cliff, the *ice barrier*. The top of this barrier averages 50 meters above the waterline. The Ross Ice Shelf itself covers over half a million square kilometers of the sea. Large sections occasionally break off to form ice islands which, because of their flat tops, are called *tabular icebergs*. Some have been observed which are 300 kilometers in length.

The glacier features of Antarctica are all on a scale befitting this principal reservoir of glacier ice on the earth. Many of these features have been presented previously for the smaller glaciers and ice fields of North America. The concluding photographs in this book illustrate a few of these features in Antarctica. Figure 128 shows a mountain range from which alpine glaciers coalesce to form valley ice streams. These streams then join to form a broad river of ice, which, in turn, flows into the featureless expanse of the Antarctic Ice Sheet. The sheet itself is also an accumulation zone, in contrast to such a temperate glacier as the Malaspina (Fig. 51), which is predominantly an ablation area with extensive melting.

Figure 129 illustrates antarctic nunataks rising out of the ice sheet. Again the scale is enormous. Many dozens of kilometers separate the nunataks one from the other, and each nunatak itself is a whole mountain range instead of an isolated peak. Figure 130 shows one of the outlet glaciers draining ice from the central antarctic plateau. Vast though this river of ice may appear, by antarctic standards it is but one among hundreds, and a small one at that; a photograph of a large one would show an undulating ice plain stretching to the distant horizon. In the background of Figure 131, several nunataks are scattered in a sea of ice, which flows toward the camera through a distant zone of seracs. The rippled surface pattern to the left and the fractured area leading to the lower right are generated by the obstacle of a buried mountain which just fails to rise above the surface and become a nunatak.

So concludes our short survey of the features of glacier ice. Of the world's wonders, abundant life-sustaining water is without doubt its most unique feature. The oceans cradled life; atmospheric circulation brings moisture to the land upon which many other forms of life developed. As brooks, rivers, and ice, water has modified the landscape into a remarkable diversity unique to this planet. Held in a delicate balance between the boiling and freezing points of water, this world is a habitable spot in a hostile universe. The climate that sustains life as we know it today lies closer to the freezing point of water, and much of modern civilization exists by virtue of a delicate balance between this climate and present snow and ice masses. Were the polar ice caps to melt, much of the world's urban area would be flooded by the rising sea. Return of another ice age would be equally destructive to the temperate zones where modern industrial society is centered.

Scientific study of glaciers, still in its infancy, has already told us much regarding the past geological and climatological history of the world. Present and future growth and shrinkage of glaciers will provide even more useful information, for these changes will be taking place when even the most subtle modifications in climate can be documented as they occur. As the amount of such information increases, the kinds of climates that led to past periods of major glaciation and their subsequent waning can be much better understood. Such understanding is essential if we are to gain a clear concept of the earth-wide environment and our place in it, for we are—in the geological sense—still in the midst of an ice age that has seen four periods of major glaciation and four periods of glacier retreat, the latter extending to the present. We do not yet know enough about the causes of world climatic changes to be able to tell whether this long sequence of Pleistocene glaciations has come at last to an end, or whether a fifth ice age still awaits us some time in the future. The answers to this question are hidden somewhere in glacier ice.

128. *Ellen Glacier, Sentinel Range, and continental ice sheet, Antarctica*

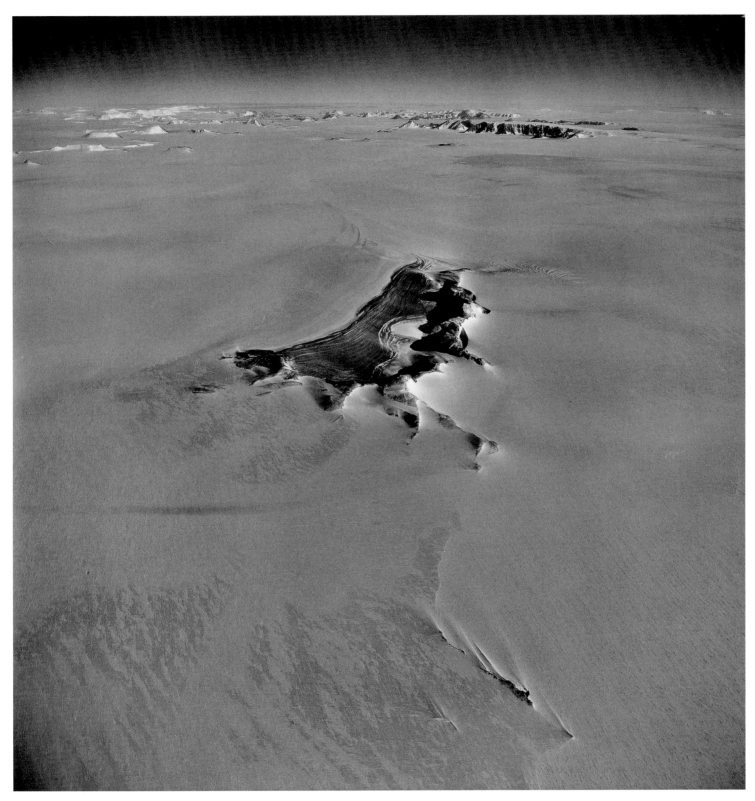

129. Mountain ranges rising above the surface of an ice sheet as nunataks, Antarctica

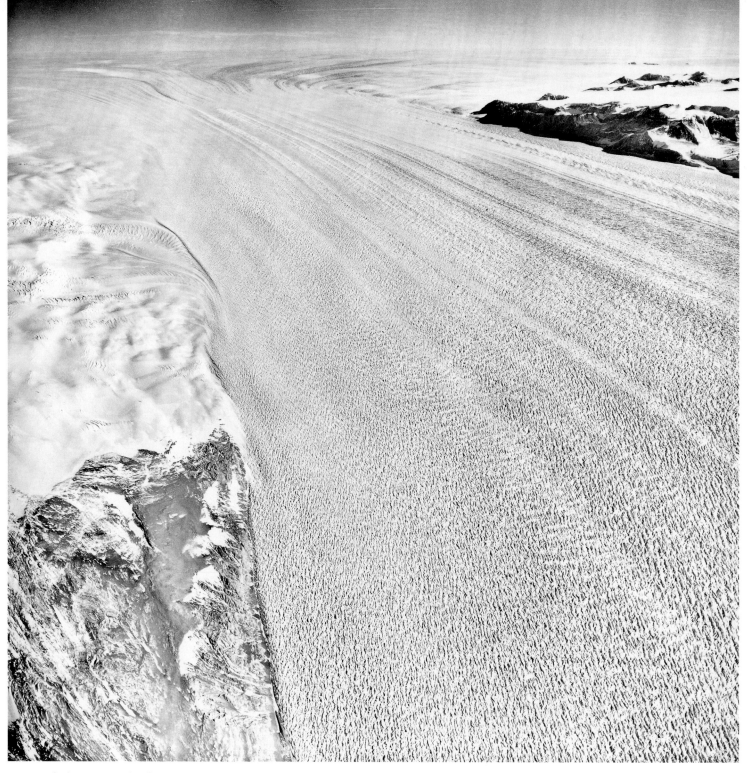

130. *Byrd Glacier, an outlet glacier from the Antarctic ice sheet draining into Barnes Inlet, Antarctica*

131. *(opposite) Flow features of the Antarctic Ice Sheet, Palmer Land*

Glossary of Glacier Terms

The numbers in parentheses give the figure numbers in this book where good examples of the described features can be found. Only the best examples are listed; similar features can often be found in other figures as well.

Ablation. The loss of mass from a glacier, usually as melt and runoff of the meltwater, but also by wind-drifting of snow or the calving of icebergs.

Ablation area. The portion of a glacier situated below the firn line where ice and snow melt exceeds snow accumulation.

Ablation moraine. Unorganized debris on a glacier surface. (47, 89, 91)

Ablation surface. A glacier surface, usually snow or firn, which has experienced appreciable melt, leading to concentration of any dirt present.

Accumulation area. The portion of a glacier situated above the firn line where snow accumulation exceeds melting.

Accumulation Area Ratio (AAR). The area of the accumulation at the end of the summer, divided by the area of the whole glacier. The higher the AAR, the healthier the glacier.

Accumulation layering. The layering in snow, firn, or ice generated by annual cycles of accumulation and ablation (analogous to sedimentary layering in rock). (9–11)

Activity index. The rate of increase of net accumulation with altitude, or the rate of decrease of net ablation with altitude.

Annual mass balance. The annual difference between net accumulation gains and net ablation losses of a glacier. A positive mass balance leads to glacier growth; a negative balance leads eventually to retreat.

Arête. A sharp and sometimes jagged ridge formed by the intersection of two cirques. (108–110)

Bergschrund. Large, semipermanent crevasse at the head of a glacier accumulation zone which separates actively flowing ice from stagnant ice above. (17, 42)

Biscuit-board topography. An upland plateau partially eroded by glacial cirques. (107)

Brash. Small fragments of floating glacier ice released by a calving glacier. This ice may be dangerous to yachts and small commercial vessels.

Calving glacier. A glacier which ends in a lake or salt water and releases brash ice and also icebergs if the glacier is large, active, or rapidly retreating. (92–94, 96, 97, 124–127)

Calving glacier advance. The process whereby a calving glacier advances in deep water by constructing a terminal moraine shoal, which reduces the calving rate. (92 [Johns Hopkins Glacier], 96)

Calving glacier cycling. Repeated calving glacier advances and drastic retreats.

Candle ice. Lake ice that disaggregates on melting into large, columnar crystals. (1)

Channel configuration. The length, width, depth, and irregularities of a channel, all of which influence the calving rate of a glacier. (92, 94, 95, 96, 97)

Chevron crevasses. A series of crevasses along a glacier margin oriented at an acute angle with the margin, usually produced by a combination of compressing flow and side friction. (21)

Cirque. A glacier-carved basin, usually with a steep headwall. (107, 109)

Compressing flow. The flow of glacier ice wherein the velocity seen at the surface decreases downstream. This is the normal state in a glacier ablation zone. (13, 32)

Crevasse. A large crack in the surface of a glacier produced by the stresses of glacier flow. (23, 79–82)

Drastic retreat. Very rapid recession of a temperate calving glacier due to receding from a shoal into deep water, which greatly increases the calving rate. (94b, 97)

Drumlin. Smoothly rounded mounds, usually of till deposited underneath a glacier and elongated parallel to glacier flow. (115)

Equilibrium Line Altitude (ELA). The average altitude of the line, or zone, dividing a glacier's accumulation area from its ablation area.

Esker. A narrow, winding ridge of till and outwash deposited by a subsurface meltwater stream of a former glacier. (117)

Extended position. Iceberg calving glacier extended so as to fill a deep inlet or fjord and terminating on a shoal, generally a terminal moraine. (94a)

Extending flow. The flow of glacier ice wherein the velocity seen at the surface increases downstream. This is the normal state in a glacier accumulation zone. (17, 18, 23)

Fiord (see fjord)

Firn. Snow in the state of transition to glacier ice, defined for convenience as snow that has survived one or more seasons of ablation.

Firn edge. The line on a glacier surface dividing firn from glacier ice, or dividing two firn surfaces of different age. (5–7)

Firn line. The average zone of maximum retreat upglacier reached by the snow line each year. The firn line separates the accumulation area from the ablation area.

Fjord. Frequently spelled fiord in Alaska. Both usages are officially recognized. Glacier-eroded valleys that have been inundated by the sea. Fjords may contain existing glaciers. (95, 96, 97, 124, 125)

Foliation. The layered structure of glacier ice generated by flow deformation and recrystallization. (83)

Glacier flour (also *glacier milk*). The extremely fine rock debris that gives a milky color to rivers and a chocolate to green color to lakes fed by glaciers.

Glacier mill (see *moulin*). (75)

Glacier sole. Debris-laden layer of ice at the bottom of a glacier. (28–31)

Glacier table. A rock left perched on an ice pillar by ablation of the pedestal and the surrounding unprotected ice. (83)

Ground moraine. Rock debris deposited underneath a glacier. (35, 115)

Hanging glacier. A glacier, usually small, clinging to a steep mountainside. Also any glacier that terminates abruptly at the top of a cliff. (12, 33, 42)

Horn. A sharp peak of resistant rock left after the surrounding landscape has been eroded away by glaciers. A horn normally marks the intersection of three or more arêtes. Its faces are the headwalls of cirques. (111)

Ice barrier. The front of an ice shelf facing on the sea. Usually a vertical cliff left behind by the calving of tabular icebergs. (127)

Iceberg (see also *ice island*). A piece of free-floating ice large enough to be a hazard to ships broken off from a calving glacier. (93, 95, 97)

Icecap. Completely ice-covered area on land, generally in the form of a dome with ice flow radiating outward from the center. (119, 124)

Ice dolines. Glacier surface features caused by drainage of sub-surface water-filled cavities and subsequent collapse of the cavity roof.

Icefall. A steep reach of a glacier with a chaotic crevassed surface and rapid rate of flow. (24, 25)

Ice field. An extensive area of interconnected valley glaciers from which the higher peaks rise as nunataks. (112, 113, 119, 126)

Ice island (also *tabular iceberg*). A very large, flat-topped section of an ice shelf which has broken off and drifted out to sea.

Ice sheet. A vast, thick layer of ice which inundates a large area of the earth's surface, i.e., Antarctic Ice Sheet, Greenland Ice Sheet. (128–131)

Ice shelf. A fixed, floating glacier which extends out from shore over a large area of the sea. (126, 127)

Ice ship. A large pinnacle of clean ice on a debris-covered glacier surface. (89–91)

Ice worm. A small annelid worm which spends its entire life cycle on snow or ice. (84)

Jökulhlaup. A sudden and sometimes catastrophic flow of water discharged from a glacier.

Lacuna. Irregular surface pockets or basins found in the ablation zone of some glaciers with a history of surges. (77, 78)

Lateral moraine. A ridge of rock debris along the side of a valley glacier. (33, 35, 45, 46, 104)

Medial moraine. A long strip of debris on the glacier surface, usually parallel to flow, which originates at the juncture of two glaciers. (48–50, 53)

Moraine. A deposit of rock debris shaped by glacial flow and erosion. (16, 33, 45–60, 78, 94, 104)

Moulin (*glacier mill*). Vertical hole by which a surface melt-water stream enters a glacier. (75)

Nieve penitentes. A field of snow or firn pillars produced by an advanced stage of sun cup development. (87, 88)

Nunatak. Land, generally a mountain peak, completely surrounded by glacier ice. (112, 113, 129)

Ogive. An arcuate band or undulation on the surface of a glacier, convex downstream, usually recurring in a periodic pattern. (61–72)

Piedmont glacier. A glacier formed at the foot of mountains by the discharge of ice from one or more confined valley glaciers. (51, 52, 120)

Pinning points. Constrictions in a channel or fjord, such as islands, headlands, points, or sharp turns, where a drastically retreating glacier may maintain tenuous stability, in some cases for several decades. (92, 94, 97)

Polar glacier. A glacier that has subfreezing temperatures throughout its mass. (123, 127–131)

Randkluft. A crack or gap between the edge of a glacier and the adjacent rock wall.

Regelation. Thawing and refreezing of ice through application and release of pressure. (29)

Regelation spicules. Small shards or feathers of ice produced at a glacier sole by the regelation process involved in the sliding of the ice over bedrock. (29)

Regional snowline. The altitude at which annual accumulation balances ablation on a ground surface.

Receiving area. The portion of a surging glacier where ice is discharged during surges, generally located in the lowest portion of the glacier. (53, 54, 57, 58, 59)

Reservoir area. The portion of a surging glacier in which ice accumulates between surges over a period of years, either by snow accumulation or by normal flow from further up the glacier. (55, 56)

Retracted position. Iceberg calving glacier that ends at the head of an inlet or fjord. (92 [Gilman Glacier])

Sea ice (pack ice). A layer of ice derived from salt water frozen on the surface of the sea (not to be confused with *iceberg, ice island, tabular iceberg*). (124–27)

Serac. A block tower or pinnacle of ice or firn formed by the intersection of crevasses. (12, 22, 23)

Snow swamp. A poorly drained area of snow or firn which is saturated with water. (76)

Snowline. The transient lower margin of winter snow accumulation on a glacier. (4, 13, 35)

Soil polygons. Patterns in the ground, sometimes several meters in diameter, produced by subsurface ice in permafrost areas. (7)

Striations (also *striae*). Shallow scratches and grooves carved in bedrock by glacier flow. (105)

Subpolar glacier. A glacier with subfreezing temperatures in most of its mass. (119, 120, 121)

Sun cups. Cuspate hollows in a snow or firn surface formed by complex ablation processes during sunny weather. (85, 86)

Surge. Periodic, very rapid movement of large quantities of ice in a glacier (active phase), alternating with long periods of near stagnation (passive stage). (37–39, 52–60, 74, 77, 78)

Tabular iceberg. An iceberg created when a large section of an ice shelf breaks away. Tabular icebergs have a horizontal dimension very large in comparison to their thickness. They are sometimes called ice islands. (125)

Tarn. A small lake occupying a hollow eroded by glacial erosion. (35, 107)

Temperate glacier. A glacier that is at the pressure melting point temperature of ice in most of its mass. (All glaciers shown in this book are presumed to be temperate except those noted as "Polar" or "Subpolar.")

Terminal moraine. A ridge of debris in front of a glacier terminus marking its maximum stand or position of readvance. (16, 35, 118)

Terminus. The lower end, or snout, of a glacier. (21, 32, 35, 38, 43, 44)

Tidewater glacier. A glacier that terminates in the sea and discharges brash and icebergs. (94–96, 124–27, 130)

Till. Unsorted rock debris deposited by a glacier. (116)

Trimline. The limit of erosion or modification of vegetation by a glacier at a position of former advance. (34, 35)

Unconformity. A discontinuity in accumulation layering. (11)

Selected Bibliography

This list of additional sources of information about glaciers and their effects on landscape serves as an introduction for the reader who seeks greater depth. The scientific literature on the subject is extensive, and books are many. We annotate here, according to level of difficulty, several introductory or textbook titles, each of which offers in turn a more extensive bibliography.

Living Ice: Understanding Glaciers and Glaciation, by Robert P. Sharp, Cambridge University Press, New York, 1988.

An excellent overview of glaciology, well illustrated and in a style easily accessible to the layperson. A good place to start.

Glaciers, by Michael J. Hambrey and Jurg Alean, Cambridge University Press, New York, 1992.

A modern successor to *Glacier Ice,* also a good place to start. A general overview of glaciology, illustrated at length with excellent color photographs.

Glaciers, by Michael George, Creative Education, Mankato, Minn., 1991.

Glossary as an art form. A brief book and an even briefer text define glaciological terms with the aid of stunning color photos and layout.

Glacier, by R. H. Bailey, Planet Earth Series, Time-Life Books, 1982.

A general introduction to the subject, illustrated with many fine photographs and diagrams typical of this series.

Glaciers and Glaciation, by Douglas I. Benn and David J. A. Evans, John Wiley and Sons, New York, 1998.

A first-rate, moderately technical text, abundantly illustrated, with emphasis on glacial geomorphology. A very thorough treatment of the subject. No fewer than eighty-five pages of references open gateways to the entire subject of glaciers.

Glacial Environments, by Michael J. Hambrey, University of British Columbia Press, Vancouver, BC, 1994.

A detailed treatment of glacial effects on the landscape (glacial geomorphology), well illustrated, moderately technical but non-mathematical. The chapter on glacier dynamics is a good summary of glacier characteristics and behavior.

The Ice Age World, by Bjorn G. Andersen and Harold W. Borns, Jr., Scandinavian University Press, Oslo, 1994.

A comprehensive picture of the most recent (Pleistocene) Ice Ages and their relationship to glacier behavior and climate. Abundant and informative illustrations.

Glacial Geology: Ice Sheets and Landforms, by Matthew R. Bennett and Neil F. Glasser, John Wiley and Sons, New York, 1996.

A comprehensive but non-mathematical overview of glaciers and their effects on landscape.

The Winters of the World: Earth under the Ice Ages, ed. Brian S. John, John Wiley and Sons, New York, 1979.

Individual chapters by various authorities cover the successive Ice Ages throughout the Earth's history and the records they have left on the landscape. Many good illustrations.

Glaciers of North America: A Field Guide, by S. A. Ferguson, Fulcrum Pub., Golden, Colorado, 1992.

Switzerland and Her Glaciers: From the Ice Age to the Present, by P. Kasser and W. Haberli, published by Kuemmerly and Frey for the Swiss National Tourist Office, Zurich, 1981.

The Greenland Ice Cap, by Borge Fristrup, University of Washington Press, Seattle, 1966.

These three books offer regional perspectives on glaciers. Ferguson covers basic glaciology and a practical guide to visiting and traveling on glaciers, often with a humorous touch. Kasser and Haberli thoroughly document, with abundant illustrations, past and present glacier conditions in a country where they have been studied longer and better than just about anywhere else in the world. Fristrup, a veteran of numerous expeditions, recounts the exploration history of Greenland's "Inland Ice," graphically depicting the character of a major ice sheet.

The Physics of Glaciers, 3d ed., by W. S. B. Paterson, Pergamon, New York, 1994.

This is a basic scientific reference on glaciers for glaciologists and other technical readers. For the experienced student or amateur scientist this text also cannot be too highly recommended. Paterson dives deeper into the subject than the non-mathematically inclined reader may find comfortable, but a discriminating reader can glean much that is important.

Satellite Image Atlas of Glaciers of the World, edited by Richard W. Williams, Jr., and Jane G. Ferrigno, U.S. Geological Survey Professional Paper 1386-G. Washington, D.C.: GPO, 1988–

The Atlas is published in eleven separate papers: (1) General discussion; (2) Antarctica (1988); (3) Asia, including India, China, Nepal, Afghanistan, Pakistan, and European Russia; (4) Continental Europe, except European Russia (1993); (5) Greenland (1995); (6) Iceland; (7) Irian Jaya, Indonesia, and New Zealand (1989); (8) Middle East and Africa (1991); (9) North America; (10) South America (1988); (11) Related geological topics.

Scientific papers in the field of glaciology abound, most of them fairly technical and often mathematical. The premier source is the *Journal of Glaciology,* published by the International Glaciological Society in Cambridge, England. The Society also publishes the *Annals of Glaciology,* proceedings of conferences on various specialized topics. *Zeitschrift für Gletscherkunde und Glazialgeologie* is another journal dedicated to the same subject. Papers on glacier-related topics appear from time to time in the *Journal of Geophysical Research,* conference proceedings of the International Association of Scientific Hydrology, *Science,* and other international journals.

Finally, for the on-line seeker of glacier information, sources are abundant, many with extensive graphics and photographs. As of late 1998, the AOL search engine reported 1,635 documents or sites related to glaciers. The exploration options are wide open. Here are a few Web sites worth visiting. There are many others as well.

http://www.glacier.rice.edu

A general, well illustrated introduction to various glacier and ice topics, with emphasis on Antarctica. Good links to allied topics such as weather (check the latest weather reports from stations around Antarctica). Much support material for teachers, along with lists of maps, books, and videos.

http://www.spri.cam.ac.uk/igs/home.Htm

Website of the International Glaciological Society. General information on publications, meetings, and Society affairs. A long list of links to other glacier-related sites worldwide.

http://wwwflag.wr.usgs.gov/GLIMS/

A site under development and well worth following. It will report data and images from the GLIMS project (Global Land Ice Monitoring from Space). Launch of the EOS-AM1 satellite to support this project is scheduled at the end of 1998.

http://hera.ethz.ch/pub_eng/dpt/glaciology/home.html

A prolific source of information, this is the Website for glaciology at the Swiss Federal Institute of Technology in Zürich. Weekly talks on glaciology plus current research projects are posted. Extensive links to relevant sites throughout Switzerland (check for the current avalanche forecast for the Swiss Alps). Included is a good list of glacier Websites around the world. The home page is in English; some of the other Swiss links are in German.

http://www.geo.unizh.ch/wgms/

The World Glacier Monitoring Service maintained at the University of Zürich in Switzerland. If you want to find the details of past and current glacier mass balances, for the world in general or for a particular glacier, this is the site to visit. Data access is moderately difficult to navigate, but it all runs in English.

PHOTOGRAPHIC CREDITS

Austin Post: Frontispiece, Figures 6b, 56, 117; *University of Washington,* Figures 4, 5, 13, 14, 16, 17, 18, 20, 21, 24, 25, 32, 34, 35, 36, 37, 38, 39, 40a and b, 41a and b, 43, 45, 46, 47, 48, 49, 51, 54, 57, 58, 59b, 61, 63, 64, 65, 66, 67, 68, 69, 70a and b, 71, 72, 74, 77, 78, 92, 94a, 95, 96, 97, 98, 99, 100, 102, 103, 107, 108, 109, 110, 111, 113, 114, 115, 118; *U.S. Geological Survey,* Figures 23, 50, 101b, 104, 106; *Polar Continental Shelf Project of the Canadian Department of Energy, Mines. and Resources,* Figures 120, 121, 122, 124, 125, 126. Flights over Northwest Greenland were made with the kind permission of the Danish government.

Edward R. LaChapelle: Figures 1, 2, 3, 11, 15, 19, 22, 27, 28, 33, 44, 73, 76, 79, 80, 82, 84, 85, 86, 105, 112, 123

Fritz Mueller, copyright Swiss Foundation for Alpine Research: Figures 9, 10, 12, 26, 29, 30, 31, 89, 90, 91 116

Bradford Washburn: Figures 53, 55, 59a, 93

A. Roch: Figures 42b, 62, 75

L. Lliboutry: Figures 87, 88

Robert Krimmel, U.S. Geological Survey: Figures 94b, 101a

M. Roch: Figure 42a

N. Untersteiner: Figure 81

Erich Vanis: Figure 42c

U.S. Navy for U.S. Geological Survey: Figures 127, 128, 129, 130, 131

National Air Photographic Library, Surveys and Mapping Branch, Department of Energy, Mines and Resources, Canada: Figures 7, 119 (A 16309 25, T 412 L 112)

Swiss National Tourist Office: Figure 83